Steam Plant

Papers presented at the one-day Seminar *Optimization and Development of Existing Steam Plant by Modelling*, held at the IMechE Headquarters, London, UK, on 1 December 1998.

IMechE
Seminar Publication

I MECH E

Steam Plant
Optimization and Development by Modelling

Organized by
The Steam Plant Committee of the Power Industries Division of
The Institution of Mechanical Engineers

IMechE Seminar Publication 1999–6

Professional
Engineering
Publishing

Published by Professional Engineering Publishing Limited for The Institution of Mechanical Engineers, Bury St Edmunds and London, UK.

First Published 1999

ISSN 1357-9193
ISBN 1 86058 234 6

A CIP catalogue record for this book is available from the British Library.

Printed and bound in Great Britain by Bookcraft (Bath) Limited.

Contents

Related Titles of Interest

Title	Editor/Author	ISBN
Handbook of Mechanical Works Inspection	Clifford Matthews	1 86058 047 5
IMechE Engineers' Data Book	Clifford Matthews	1 86058 175 7
Steam Turbine Governing and Overspeed Protection	IMechE Seminar 1998–10	1 86058 162 5
Remanent Life Prediction	IMechE Seminar 1998–4	1 86058 154 4
Gas Explosions in CCGT and Steam Plants	IMechE Seminar 1998–3	1 86058 169 2
Hydro Power Developments – Current Projects, Rehabilitation and Power Recovery	IMechE Seminar 1997–4	1 86058 121 8

For the full range of titles published by Professional Engineering Publishing contact:

Sales Department
Professional Engineering Publishing Limited
Northgate Avenue
Bury St Edmunds
Suffolk
IP32 6BW
UK

Tel: 01284 724384
Fax: 01284 718692

S593/001/98

Conversion options for Peterhead Power Station

I R DIBBLE and **I H D O'DONNELL**
Scottish Hydro-Electric plc, Perth, UK
G F HUNT
Thermoflow Limited, Tewkesbury, UK

SYNOPSIS

The two gas/oil-fired 660 MW generating sets at Scottish Hydro-Electric's Peterhead Power Station were originally commissioned in 1981/2 and in order to remain competitive it was essential to upgrade the plant. This paper describes the plant performance aspects of a study carried out by Scottish Hydro-Electric to identify the possible routes for plant modification and the associated efficiency of each option.
Scottish Hydro-Electric employed the suite of Thermoflow programs to carry out the performance study. These programs had sufficient flexibility to model the performance of the original plant and then to analyse the variety of re-powering options available, using the original accurate model as the basis.

1. INTRODUCTION

1.1 Introduction

The current generation of combined cycle power plants offer efficiencies now approaching 60%. Existing power plant based on subcritical Rankine cycles have efficiencies in the mid to high 30's. In order to maximise the asset value of existing stations, either a cheaper fuel or higher efficiency is essential if such plants are to remain competitive and environmentally compatible. Scottish Hydro Electric resolved this problem at its Peterhead Power Station by taking the decision to repower one of its two gas/oil fired 660MW steam units. This decision was taken in light of the studies described here to investigate the re-powering options available.

1.2 Existing Station

The reheat units at Peterhead have steam conditions of 160 bar, 538°C, 538°C and an efficiency of approximately 38%. The plant was designed and built by the predecessors of Mitsui Babcock and Alstom Energy. The boiler, which was designed to burn a wide range of gases, natural gas liquids in the vapour phase and heavy fuel oil, is of the El Paso natural circulation type with ribbed tube furnace water walls and horizontal superheat and reheat tubebanks. The steam turbine was the first UK 660

MW unit to be based on 4 cylinders rather than the 5 previously favoured. Additionally there are two GE Frame 9E 115MW open cycle gas turbines. These were commissioned in 1992.

1.3 Reasons for Re-powering

The units at Peterhead have approximately two thirds of their original design life remaining and needed enhancement if the plant's full asset value was to be realised. Medium term fuel forecasts suggest that natural gas will remain the competitive fuel. Since other UK generators, with higher efficiency plants, will also be burning natural gas it is essential for Peterhead to generate at comparable levels of efficiency in order to remain competitive. By re-powering, efficiency levels and environmental performance can be increased to match combined cycle plant at approximately 60% of the cost of a greenfield combined cycle power plant. This would ensure that the asset's utilisation could be maximised to provide an attractive return with minimised capital expenditure.

2. OPTIONS FOR CONVERSION

Four main re-powering options were considered:

- Hot Windbox
- Feedheating
- Combined Cycle
- Hybrid

2.1 Hot Windbox

Hot Windbox re-powering utilises the exhaust from a gas turbine to replace or supplement the existing combustion air supply to the existing fired boiler.
The hot vitiated gas turbine exhaust passes more heat directly into the fired boiler furnace than the existing plant's draught system. This results in less fuel being required in the fired boiler to produce the same total furnace heat input and steam conditions as the original plant.

A typical hot windbox re-powering cycle diagram is shown in Figure 1 and in (1), (2).

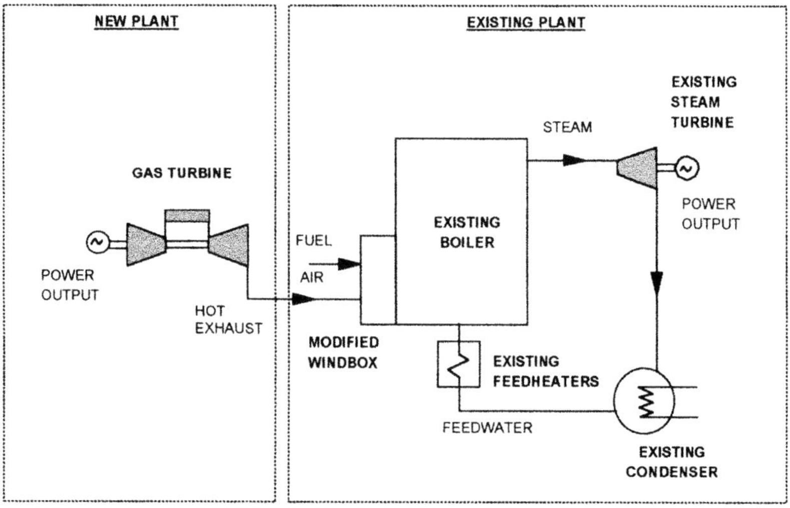

Fig. 1 Hot Windbox Re-powering Cycle Diagram

The Hot Windbox schemes considered for Peterhead were:

- Utilising the existing Frame 9E gas turbines
- New gas turbines

2.2 Feedheating

Feedheating re-powering utilises a gas turbine's exhaust to supplement or replace an existing plant's feedheating by a heat exchanger in the gas turbine exhaust.
The existing plant's fired boiler is still used to raise steam to the required conditions.
The benefit of this scheme is to reduce the amount of steam being bled off the steam turbine for feedheating, which for a plant like Peterhead amounts to as much as 30% of the steam input to the turbine. This increases the steam turbine output for the same boiler fuel input. A typical feedheating re-powering cycle diagram is shown in Figure 2 and in (2), (3).

The existing Frame 9E gas turbines at Peterhead were a good fit for this scheme and therefore this option was modelled using them.

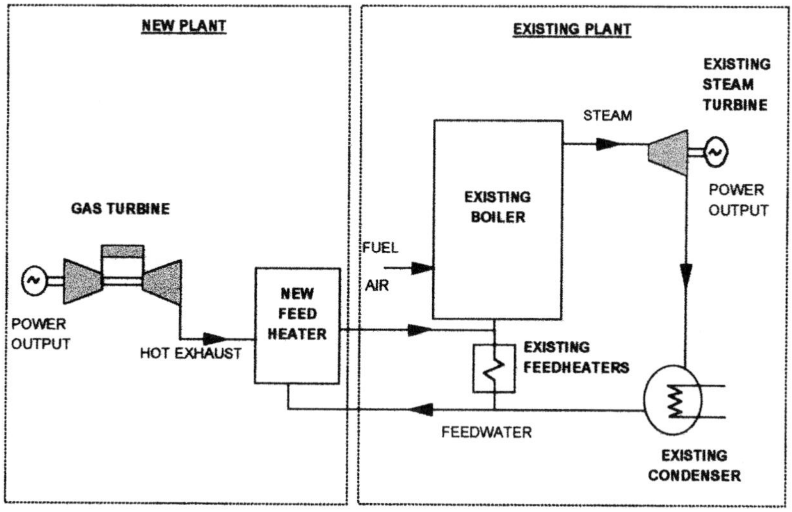

Fig. 2 Feedheating Re-powering Cycle Diagram

2.3 Combined Cycle

Combined cycle re-powering utilises a new steam source, in this case gas turbines with heat recovery boilers, to replace the existing fired boilers steam production.
The open cycle gas turbine efficiency is enhanced by the generation of steam in a heat recovery boiler, which is fed to the existing steam turbine to give a considerably better efficiency than the original steam plant. A typical combined cycle re-powering cycle diagram is shown in Figure 3 and in (2).

To enable combined cycle operation of the plant, the new gas turbines had to be sufficiently large to provide sufficient quantities of steam to power the existing steam turbine. Typically the recovery from a gas turbine's exhaust is equivalent to half of the gas turbine output. The largest gas turbines available have an output of >250MW and therefore, for each gas turbine, approximately 130MW of steam turbine output would be available. A balance between having enough steam to enable the steam turbine to operate and achieving full load from the steam turbine had to be struck.

Two gas turbines would enable the existing steam turbine to be operated in combined cycle mode, but if one had a planned or forced outage, the steam production from the other would not be enough to meet the steam turbines minimum load point. Converselv achieving full load from the steam turbine would require up to five gas turbines. The output from the combined cycle plant would be in excess of the transmission system capacity and the desire to have the re-powered station's load factor as high as possible would therefore be affected.

Consequently this case was modelled using three or four of the largest, most efficient gas turbines available.

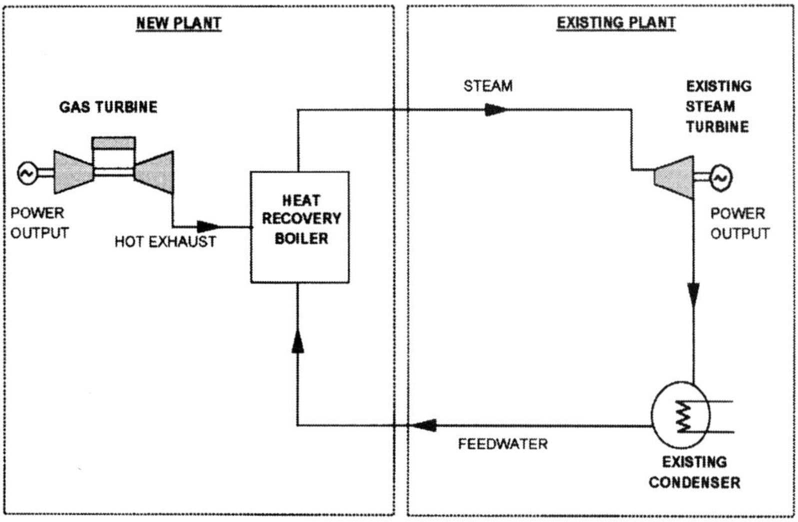

Fig. 3 Combined Cycle Re-powering Cycle Diagram

2.4 Hybrid

Hybrid re-powering, as with the combined cycle re-powering, utilises a new steam source but retains the existing fired boiler. The benefits are the same as for the combined cycle re-powering but are dependant upon the level of existing fired boiler capacity that is used. The main benefit of hybrid re-powering is that the plant flexibility and fuel options are better than the combined-cycle mode due to the retention of the existing boiler. If the gas turbines are large enough, the plant could be switched between combined cycle and hybrid mode operation as desired. A typical hybrid re-powering cycle diagram is shown in Figure 4.

The hybrid schemes considered were:

- Use of one or both of the existing Frame 9E gas turbines. These gas turbines would not be able to supply sufficient quantities of steam to enable combined cycle operation.
- Use of the largest available gas turbines to enable hybrid or combined cycle operation.

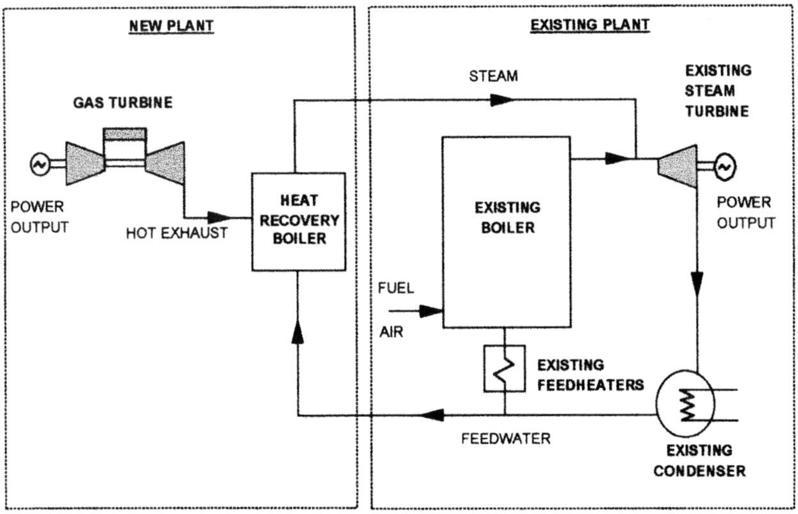

Fig. 4 Hybrid Re-powering Cycle Diagram

3. PROGRAM CHOICE AND CONFIGURATION

3.1 Considerations Leading to Choice of Thermoflow Suite of Programs

Scottish Hydro-Electric had previously used Thermoflow's GTPRO and GTMASTER programs for studies on gas turbine plant in combined cycle and combined heat and power (CHP) schemes. GTPRO is now firmly established in this field, with approximately 1000 users all over the world. When the re-powering of Peterhead was first mooted, Thermoflow had already expanded their suite of programs to include modelling of conventional steam plant using programs called STEAM PRO and STEAM MASTER. A further program called RE-MASTER had also just been introduced to combine the detailed design and simulation capabilities of GTPRO, GTMASTER, STEAM PRO and STEAM MASTER and aid the user through the selection and conceptual design of a re-powering scheme.

The major advantage to the user in all these programs is that the model can quickly and easily be constructed by an engineer familiar with the plant. Time does not need to be spent de-bugging and iterating towards a final solution, as would be required by a fully-flexible modelling package.

3.2 Re-powering program - General Approach

The general solution of a re-powered conventional steam plant is particularly challenging in that it requires not only a model of the existing plant, but also the ability to model new, custom designed gas turbine plant in a CHP configuration, in any of the configurations described in 2 above. The existing plant will invariably operate off-design when the re-powered plant operates at its new rated steam conditions. The integration of new and existing components requires the matching of flow rates, pressures and temperatures of every stream that passes from one part to the other.

The modelling concept created by Thermoflow uses STEAM PRO and STEAM MASTER to reproduce the characteristics of the existing plant, accurate at off-design conditions and part loads.

This forms the starting point of the re-powering conceptual design. The new plant may comprise a gas turbine and heat recovery boiler, which can be readily modelled in GTPRO and GTMASTER. New heat exchangers, contributing heat to the feed water heating system of the existing plant may be modelled in the STEAM MASTER package as additions. The re-powering program, RE-MASTER, in computing the combined plant performance co-ordinates the STEAM MASTER and GT MASTER programs to suit the chosen configuration.

Each group of programs is described in turn in the following sub-sections.

3.3 Steam Plant Modelling Programs

The scope of STEAM PRO covers gas, oil or coal-fired boilers, steam turbine plant, condenser, the condenser cooling system, feed water heating systems, process and district water heating systems. Within the standard menus, which the user considers when designing the plant, there is an enormous range of flexibility covering small back pressure turbines through to large nuclear applications.

The menu structure follows a logical progression leading from early general decisions, through to specific numerical entries necessary to define the plant. An example of one of the later, more important, menus within STEAM PRO is given in Figure 5 and is typical for all Thermoflow's application-specific programs. The program as installed will include reasonable defaults for all values in every menu. The user enters in the appropriate box the values he wishes to specify. Many inputs or indeed complete menus can be ignored for the purposes of outline design; but if the user prefers to have more control he can delve into the details.

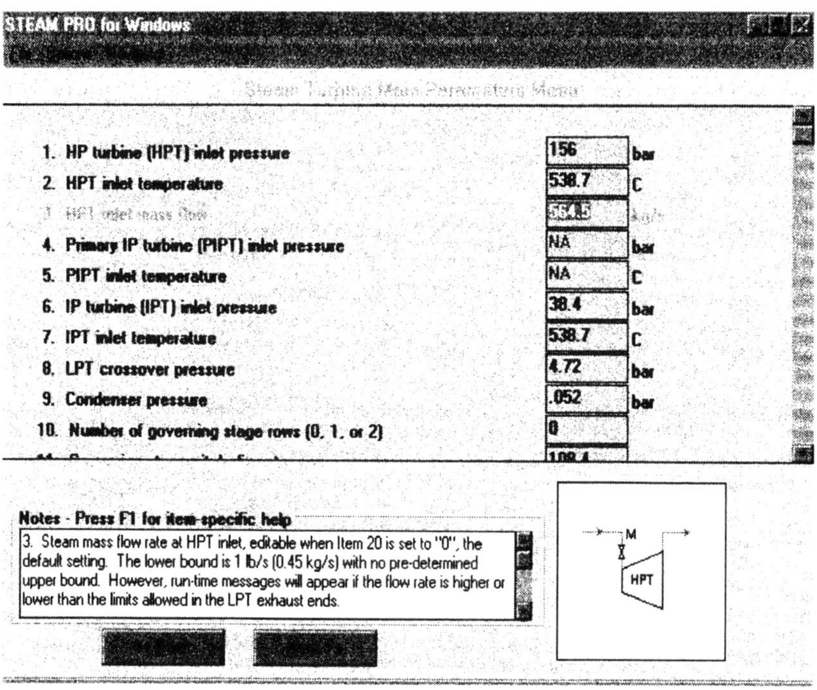

Fig. 5 STEAMPRO Menu

The text output and graphic displays give a complete representation of the plant's energy balances. Detailed tabular data is given on each component and specific graphical displays show the boiler, turbine schematic, cooling water system, boiler heat recovery diagram and turbine expansion line.

The complementary STEAM MASTER program enables the user to investigate off-design performance. The hardware designed by STEAM PRO becomes the foundation of the STEAM MASTER model. It also permits further detail in the hardware to be specified, for example parallel strings of feed water heaters or moisture extraction from wet steam at turbine bleed ports. Pressure drops, either relating to steam or water in the pipework or to the flue gas as it passes through the boiler heat exchangers, which were fixed at design values in STEAM PRO, are computed using industry-standard pressure loss calculations.

In STEAM PRO the steam turbine model is semi-empirical. Hardware information, adequate to represent the pressure-flow characteristics of the turbine, is established from physical principles. However, the turbine efficiency, steam expansion path, leakages and losses are determined by the much-followed method of Spencer, Cotton and Cannon (4).

In STEAM PRO the user has the ability to modify the efficiency as determined by Spencer, Cotton and Cannon by a correction factor, whilst in STEAM MASTER the user has full control of the values determined for efficiency and group nozzle area. If the user chooses not to modify these parameters, reasonably conservative determination of performance are made at both design and off-design conditions. Other features of steam turbine design, such as the specialised treatment of last stage blades and their leaving losses are covered by the program, but not dealt with in this paper.

3.4 Gas Turbine and HRSG Modelling Programs

GTPRO and GTMASTER are engineered on the same principles as STEAM PRO and STEAM MASTER but include a gas turbine database as the starting point for establishing a conceptual design. Since 1987 these programs have progressively developed, with new gas turbines and cycle features being added to the database as the manufacturers develop them.

Because there is a gas turbine cycle model for most of the 150 machines in the database, it is possible to make small adjustments to firing temperatures, compressor and turbine efficiencies, etc. The user can therefore easily represent the performance of detailed variants not included in the database or of the database machine when it has deteriorated due to ageing or fouling. The program enables operation at part load with a variety of control options, ranging from the manufacturers' standard configuration to one controlled by the user. The program covers numerous other add-ons for the gas turbine. These include fuel or inlet air heating and pressure boosting, combustor cooling, steam or water injection and operation from any standard or user-defined fuel.

GTPRO covers a broad range of steam systems, with or without supplementary firing, ranging from simple heat recovery boilers to multi-pressure steam-injected gas turbines and reheat combined cycles with numerous process steam or water extractions or additions. The HRSG itself may be designed automatically by the program or the user can manipulate up to 23 heat exchangers, a radiant boiler and any of two duct burners to meet a desired configuration. The steam turbine uses a similar system to STEAM PRO.

GTMASTER is the complementary simulation program to GTPRO and allows more detail to be added to plant and the design to be modified in hardware terms if required. Off design performance may then be readily computed with appropriate control systems e.g. for maintaining process pressures at part loads.

3.5 The Re-powering Program, RE-MASTER

The function of the RE-MASTER program is to guide the user through the choices and decisions involved in the re-powering process and inform him as the design proceeds of any limitations imposed by the existing plant or choices he has already made. The decision tree of the RE-MASTER program is shown in Figure 6.

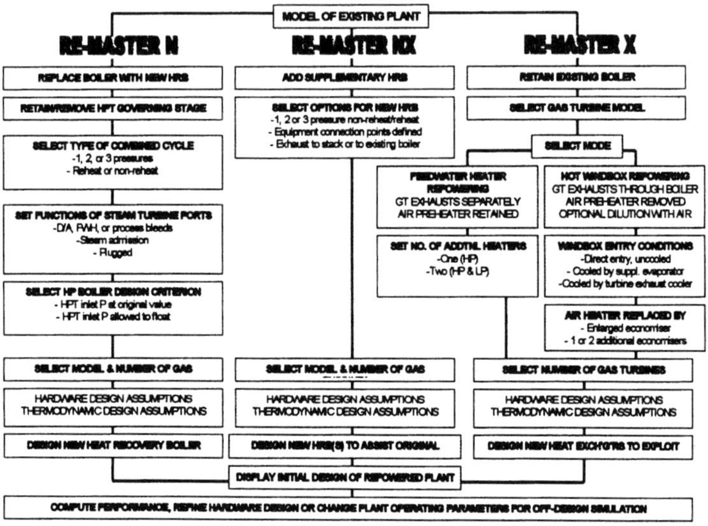

Fig. 6 RE-MASTER Decision Tree

Typically, the program reacts quickly to inform the user of thermodynamic constraints but does not know enough about the mechanical construction and metallurgical limitations of the plant, so the user must check these limitations following the achievement of a reasonable design concept.

RE-MASTER is able to accomplish the thermodynamic integration of the new plant with the existing plant and eliminates the time-consuming, manual estimation of the existing plant's off-design behaviour, that would otherwise need to be done for each iteration of the process. The model of the existing components is manipulated to rapidly converge upon the conditions which match flows, pressures and temperatures at each juncture where the existing and new equipment interact. All these processes are of only secondary concern to the user, since RE-MASTER takes care of all the interfaces with the other programs called up and co-ordinates their operation in the computation of the re-designed plant.

Once the user has secured a reasonable conceptual design, he may readily compare it in terms of thermal efficiency with any of the alternative re-powering methods. RE-MASTER gives a wealth of detail in its output text and diagrams. An example of the graphical output, showing the re-powered boiler schematic for hot windbox re-powering with two additional economisers, ECA and ECB, is shown in Figure 7.

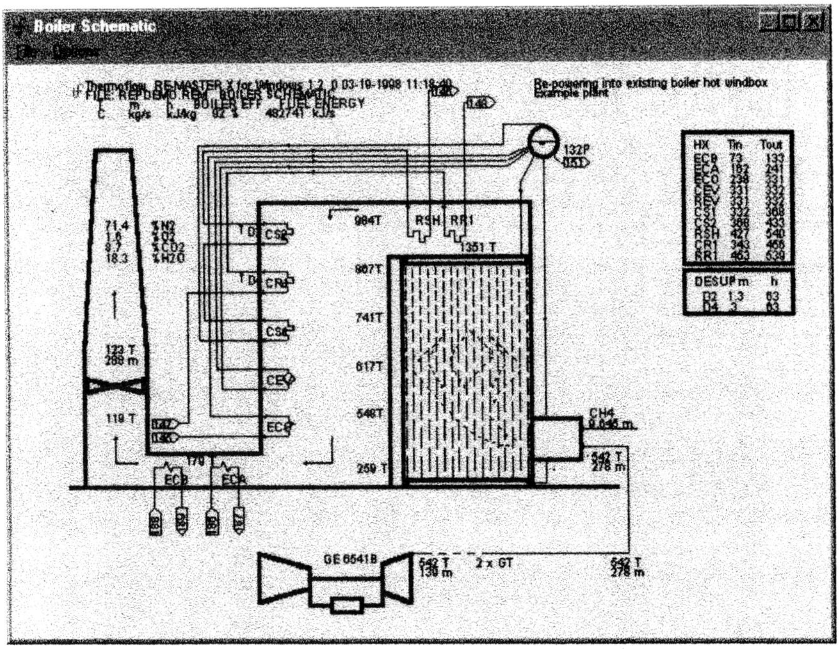

Fig. 7 Example of Graphical Output Showing a Hot Windbox Re-powered Boiler Schematic

The auxiliary power consumed by all ancillaries in the re-powered condition will be computed and the exhaust composition and mass flows from the boiler(s) will be evident. Having selected the final design concept, the preliminary designs of additional equipment such as heat exchangers may be created. Finally the behaviour of the re-powered plant may be simulated at part load or at other off-design conditions.

4. MODEL OF EXISTING PLANT

The existing plant was modelled by Scottish Hydro-Electric using STEAM PRO and STEAM MASTER.
Firstly the plant's design data was input into the model to check the program's thermodynamic calculation outputs against the as built plant performance data. The model output was found to accurately match the as built data.

The scope of data required for this included:

- Boiler

  Boiler type and furnace dimensions
  Tube bank layouts and water/steam conditions
  Fuel type
  Draught plant and air heater details
  Stack conditions

- Feedheaters

  Number, layout and type
  Inlet and outlet conditions

- Steam Turbine

  Layout, e.g. for Peterhead one HP cylinder, one double flow IP and two double flow LP cylinders
  Inlet and extraction conditions
  Condenser conditions

Secondly, part load conditions were modelled through the use of STEAM MASTER's ability to model off design and part load conditions. The ability of the program to achieve this is especially important in modelling the various re-powering options considered, as the existing plant is operated off-design and at part load.

Lastly, up to date operational data was input into the model to replicate the plant in its existing state, e.g. steam turbine degradation, feedheater performance, boiler performance and plant upgrade data since first commissioned. This formed the basis of the model to be used for modelling the re-powering options.

5. STUDY OF RE-POWERING OPTIONS

The four main re-powering options were modelled using the re-powering modelling software RE-MASTER. RE-MASTER enables the user to select the type of re-powering desired, based on an existing STEAM MASTER model. The program incorporated sufficient flexibility to enable all the options discussed in Section 2 to be studied.

Table 1 shows the performance of the re-powering options considered:

Table 1: Comparison of Re-powering Options

Case	**Existing**	**Hot Windbox**		**Feedheating**	**Combined Cycle**	**Hybrid**	
Description	Existing Plant	Existing GT's	New >250 MW GT	Existing GT's	3 x >250 MW GT's	Existing GT's	3 x >250 MW GT's
ST Output, MW	660	660	660	660	360	660	625
GT Output, MW	0	115	265	115	3 x 265	2 x 115	3 x 265
Total Output, MW	660	775	925	775	1155	890	1420
Efficiency, normalised	1	1.08	1.18	1.08	1.45	1.10	1.35
Cost, normalised	0	0.25	0.5	0.1	0.95	0.25	1

Hot Windbox with the use of the existing gas turbines had the benefit if utilising all the existing plant at the station and hence the capital cost was low. However the efficiency gains were small. Whereas with new larger gas turbines greater efficiency gains were possible but were still well below the performance of new combined cycle gas turbine plant.
Feedheating re-powering again offered the use of the existing gas turbines but the efficiency gains were small.
Combined cycle re-powering had a high capital cost but also the highest efficiency gain.

Hybrid re-powering with the existing gas turbines offered a low capital cost but also a low efficiency gain, whereas with the new large gas turbines offered a high efficiency gain, although the cost was the highest of all the options.

6. SCHEME SELECTION

Hybrid re-powering with new large gas turbines had the highest capital cost but also one of the highest efficiency gains. Additionally the plant could also be operated in the combined cycle mode, which would achieve efficiency levels close to that of the latest combined cycle power plant. A further benefit was that this option was the most flexible as combined cycle, hybrid or existing modes could be operated. A wide range of fuels could also be utilised in the overall plant.

The hybrid re-powering options, which had been studied thermodynamically, were then tested against a comprehensive range of fuel and electricity market conditions. This work using in-house software, showed that three, new, large gas turbines gave the best risk weighted market return. The high level of efficiency would future proof the plant and enable it to compete on even terms with pure CCGT plant whilst retaining the flexibility in fuel burn and operating modes.

Tenders for the construction of the re-powering scheme were submitted in October 1997 and the Thermoflow software was used in the evaluation of the tenderer heat balance submissions.

7. CONCLUSION

Scottish Hydro-Electric found the Thermoflow programs used had sufficient flexibility to model the diverse range of re-powering options examined and their ease of use enabled rapid evaluation of all the options considered.

The initial STEAM MASTER model correlated well with the existing plant's full and part load test results. The accuracy of the final RE-MASTER model is demonstrated by the fact that it closely matched the tenderer heat balance submissions. This gave confidence that the performance predictions of the other re-powering options modelled were accurate and well founded.

REFERENCES

(1) BRÜCKNER H., FINCKH H.H. AND VEENEMA J.J., Topping the Groningen Steam Turbine Plant with a Gas Turbine, ASME Paper 87-GT-38, 1987.

(2) STOLL H.G., SMITH R.W. & TOMLINSON L.O., Performance and Economic Considerations of Repowering Steam Power Plants, GE Paper GER-3644D, 1994.

(3) CASTANIER A. & BEAU J-C, Power Plant Re-powering by Steam-Gas Combined Cycle Conversion, Alsthom Review No.10, 1988.

(4) SPENCER R.C., COTTON K.C. & CANNON C.N., A Method for Predicting the Performance of Steam Turbine-Generators 16500kW and Larger, GE Paper GER-2007C, July 1974.

S593/002/98

Optimization of power station performance and design

C R GANE
British Energy plc, Gloucester, UK
B SCRUTON
Consultant, Southampton, UK

SYNOPSIS

To ensure optimum design in the replanting or refurbishment of power plant components it is essential to assess components within the framework of a whole-plant model. In the British nuclear power industry the NUMEG suite of whole-plant computer models is used for this purpose. The principles and use of NUMEG are illustrated here by reference to the replanting of steam turbines in British Energy's AGR power stations.

1. INTRODUCTION

Steam turbines now available have appreciably higher efficiencies than those installed in power stations constructed ten to twenty years ago. There are therefore potential benefits in plant output to be gained by replanting turbines. To achieve the maximum benefit, it is necessary to match the new turbines to the remainder of the plant. Essentially this requires that the swallowing capacities of the new turbines should be such as to produce an appropriate pressure distribution in the steam circuit, for example to ensure a sufficient supply of bled steam to the turbine-driven boiler feed pump or to satisfy steam generator pipework pressure limits. Consideration must also be given to the possibility of plant conditions changing appreciably within the lifetime of the replanted turbines. For example in recent years a number of British Energy's Advanced Gas-Cooled Reactors (AGRs) have experienced a change in boiler performance as a result of the deposition of carbonaceous material on the gas side (primary side) of the boiler tubing. Optimization of the design of replanted turbines should ensure compatibility with all envisaged future scenarios.

Recently, the high pressure (HP) and intermediate pressure (IP) turbines associated with several of British Energy's AGR power stations have been replanted. In each case, the optimization of the design of the replanted turbines has been guided by applying the NUMEG suite of whole-plant computer models[1,2] which have been developed and extended over several years to represent all seven of British Energy's AGR power stations. The NUMEG models embody the widely used boiler code NUMEL, originally developed for monotube modelling of the steady-state and transient responses of once-through boilers[3], but also incorporate models of all the major items of plant: the reactor, gas circulator, high pressure (HP), intermediate pressure (IP) and low pressure (LP) turbines, condenser, cooling water and feed heating systems, de-aerator, boiler feed pump, boiler feed pump turbine and the various valves.

This paper gives an account of the application of NUMEG to the optimization of the design of replanted HP and IP turbines for one of the British Energy AGR power stations, illustrating the importance to this process of two key capabilities of NUMEG: the model fitting procedure (MFP) and the plant optimization procedure (POP).

2. PLANT DESCRIPTION

The arrangement of the AGR plant is shown schematically in Figure 1. Coolant gas (carbon dioxide) from the reactor is circulated downwards through the boiler units, each comprising a high-pressure (HP) boiler and a reheater. Steam generated in the HP boiler drives the HP turbine, from which the cool exhaust steam, after being reheated in the reheater tube banks, drives the IP turbine before passing to the LP turbines and condenser. Steam bled off downstream of the HP turbine, together with live steam taken off upstream of the HP turbine governor and stop valves, drives the boiler feed pump turbine, the exhaust steam then passing to the deaerator. The feed water from the condenser, after passing through the generator coolers and chemical polishing plant, is heated in the direct-contact feed heaters by steam bled from the LP turbine stages.

3. NUMEG WHOLE-PLANT MODEL

NUMEG models all the plant items depicted in Figure 1. The boiler is modelled[3] as a one-dimensional counterflow heat exchanger. On the secondary (water/steam) side, the conservation equations of mass, momentum and energy are solved for the bulk values of mass flux, pressure and enthalpy up the height of the boiler. On the primary (gas) side, only the energy conservation equation is solved in full for the gas temperatures. The one-dimensional partial differential equations which result are reduced to algebraic form by a finite-difference method and solved by a Gaussian elimination method to yield the flows, pressures, enthalpies (and corresponding temperatures) up the height of the boiler.

The other plant modules are algebraic in nature and consequently do not require the solution of differential equations. For example, the turbine equations[1] are posed in terms of an efficiency for each stage, relating the outlet enthalpy to the inlet enthalpy, and an ellipse law constant (swallowing capacity) relating the steam flow to the inlet and outlet pressures (see section 4).

The plant optimization procedure has a number of applications, including the calculation of set points (plant boundary conditions) to achieve the maximum electrical output, the optimization of replacement plant components (e.g. the design of additional or replacement indirect-contact feed heaters) and the assessment of performance gains from upgraded plant items (e.g. bled steam lines to feed heaters). There are typically around 30 plant constraints to be satisfied to ensure safe and prolonged station operation.

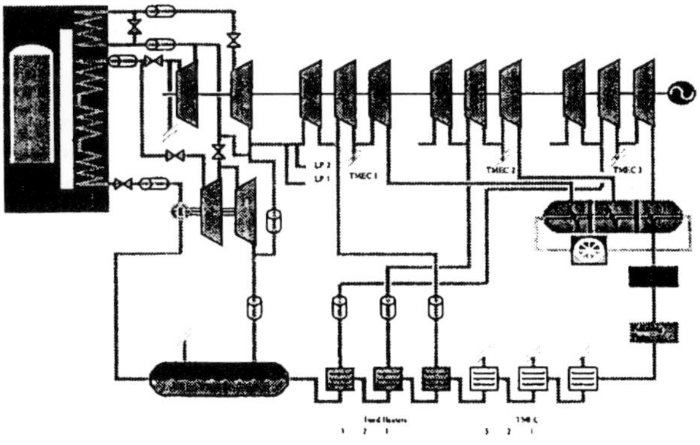

Figure 1: Plant Diagram for One of British Energy's AGRs

3.1 Model Fitting Procedure

The NUMEG codes have embodied within them the facility to adjust selected model/engineering parameters to achieve a best overall fit to a set of plant measurements, allowing for instrument uncertainties. This facility, referred to as the model fitting procedure (MFP), uses a scaled Lagrangian method with undetermined multipliers[4]. The MFP has been used in the process of validating the model against plant data. Recently it has also been used as an integral part of on-line plant condition monitors[5] where 'live' plant measurements are processed and model internal parameter values adjusted to give current best model-to-plant fits. Quantities that can be monitored in this way include fouling heat transfer resistances on the boilers and reheaters and increased flow resistance and reduced heat transfer due to any crudding inside condenser tubes.

The MFP minimizes the sum of squares of differences between model predictions (x) and plant measurements (p), each square term being weighted by the standard deviation (s) of the measurement error, i.e. the MFP minimizes an objective function E expressed as

$$E = G_1^2 + G_2^2 + \cdots \tag{1a}$$

where each G_i is of the form

$$G_i = (x_i - p_i)/s_i \tag{1b}$$

The minimization of equation 1 is obtained subject to equality constraints (residual equations):

$$f_i = 0, \qquad i = 1 \text{ to } n \tag{2}$$

where n, the number of residuals, is typically between 30 and 40. Each residual equation expresses the requirement that two variables must be identical in the final solution, e.g. the boiler inlet pressure (P_{bi}) and feed pump exit pressure (P_{fpe}). The corresponding residual f_i is the ratio of the two values minus unity (i.e. $P_{bi}/P_{fpe} - 1$ in this case), which must be less than 10^{-4} in absolute magnitude for convergence of the whole plant iteration cycle. Using Lagrange's method of undetermined multipiers (a_j), the Lagrangian L is defined as

$$L = E + \sum_j a_j f_j \tag{3}$$

This expression is differentiated to find the minimum point. This yields a set of non-linear equations, which are solved iteratively to derive the values of the independent variables y_j and the corresponding a_j, allowing for any bounds on the independent variables[4].

Any of NUMEG's independent parameters (y) may be varied in value, either freely or within bounds, to assist in the process of finding a best fit to plant data. An important example of such a parameter is the fouling heat transfer resistance associated with external carbonaceous deposition on boiler and reheater tube surfaces.

3.2 Plant Optimization Procedure

Whereas the steady state finder achieves a steady state condition for a fixed set of plant boundary conditions, the plant optimization procedure (POP)[6], linked to the NUMEG suite of programs, allows parameters (referred to as optimization parameters), that would normally have a fixed value, to be varied in an attempt to achieve some optimum steady state performance. When the optimization procedure is executed, it first calculates the steady state operating condition for a set of optimization parameters supplied by the user. A mathematical function, usually called the objective or cost function, is then used to measure how well the trial optimization parameter set meets the objectives defined by the user. The procedure then varies the values of the optimization parameters, within prescribed constraints (lower and upper bounds), to maximize the 'objective' function value while also satisfying constraints on specified NUMEG dependent variables. Most commonly the objective to be maximized is either the generator output or the sent-out electrical power.

The imposed constraints, typically numbering about 30, correspond to various flow, temperature and pressure limits, together with physical constraints such as valve openings. For each active constraint the optimization procedure gives a 'marginal cost', i.e. the change in the value of the objective for a unit change in the value of the constraint.

4. TURBINE REPLANT

4.1 Application of Plant Optimization Procedure

In using NUMEG to optimize the design of the turbines the aim was firstly to confirm that a substantial gain in electrical output would result from the replant and secondly to ensure that the optimized design would be tolerant to envisaged future changes in plant conditions. To achieve the maximum benefit, it is necessary to match the new turbines to the remainder of the plant. Essentially this requires that the swallowing capacities of the new turbines should be such as to produce an appropriate pressure distribution in the steam circuit, for example to satisfy steam generator pipework pressure limits.

The plant optimization procedure in NUMEG is ideally suited to this purpose. It enables the net plant output (generated output minus gas circulator power) to be maximized subject to the condition that all the plant constraints are satisfied, the turbine swallowing capacities being simultaneously varied to achieve their optimum values. The turbine swallowing capacity, C_{el}, is defined as

$$C_{el} = \frac{P_i v_i W^2}{P_i^2 - P_o^2} \tag{4}$$

where P_i and P_o are the inlet and outlet absolute pressures, v_i is the steam specific volume at inlet and W is the steam mass flow. The swallowing capacities of the HP and IP turbines are included in the list of optimization parameters together with other 'boundary conditions' which are allowed to vary in the optimization procedure. Typically the optimization parameters include:

Boiler feed flow
Boiler gas flow
Gas pressure
Gas temperature at inlet to the boiler unit
HP governor valve pressure drop
IP governor valve pressure drop
Pod valve pressure drop
Proportion of reheat steam diverted to reheater bypass
Flow of live steam to the boiler feed pump turbine
HP turbine swallowing capacity
IP turbine swallowing capacity

The turbine efficiencies are assigned the values specified by the manufacturer. The net plant output is designated as the objective to be maximized.

In order for the optimization procedure to be effective it is essential that input data used in the program are valid. Data for plant items other than the turbines to be replanted are assigned values resulting, where appropriate, from a recent model-to-plant fit using the model fitting procedure. Failure to achieve an adequate fit could cause misleading results since an incorrect representation of part of the plant could cause the wrong constraints to be active in the modelling. An important parameter is the heat transfer impairment (due to carbon deposition) assigned to the reheater and secondary superheater (i.e. the upper part of the HP boiler). It is modelled in NUMEG as a thermal resistance (r_d) which modifies the convective heat transfer coefficient on the primary (gas) side of the heat exchanger tubes:

$$\frac{1}{h_t} = r_d + \frac{1}{h_f} \tag{5}$$

where h_f is the heat transfer coefficient from the bulk gas to the deposit outer surface and h_t is the resulting heat transfer coefficient from the bulk gas to the tube outside metal surface. Where appropriate, account is taken of the effect of any finning on the outside of the boiler tubes. The value of the heat transfer impairment due to carbon deposition is now regularly monitored by use of the model fitting procedure 'on-line'[5].

When the turbine swallowing capacities have been established by the procedure outlined above, their values are retained unchanged in the program data and the corresponding variables are removed from the optimization parameter list. Further applications of the optimizer then enable the optimized output from the unit in its post-replant condition to be forecast for any envisaged set of circumstances. Thus it is possible to account for a variety of aspects of plant aging, such as changes in the condition of heat transfer surfaces and reductions in turbine efficiency or feed pump efficiency, in order to confirm that the gains in output from the replant, relative to output from the existing plant, are still appreciable. By suitably varying the conditions assumed in the turbine optimization, alternative scenarios can be investigated to enable a decision to be reached on whether it is preferable to optimize the turbines for current conditions or for envisaged future conditions, or for a compromise between the two.

Thus the use of the optimizer guides the design of the replacement turbines, enables the gain in output to be reliably predicted and enables the benefits of the replant to be assessed with regard to future plant conditions and operating scenarios. Furthermore, the optimization procedure provides information on the marginal cost of each active constraint and thereby enables the dominant (most power-limiting) constraints to be identified.

4.2 Confirmation of Replant Benefits

Subsequent to the turbine replant an important consideration is the confirmation that the expected benefits of the replant have been achieved. In a particular case in which both the HP and IP turbines were replanted, the turbine manufacturer and station engineers were active in monitoring the level of enhanced performance following return to power. Plant measurements were obtained for both pre-replant and post-replant conditions. Using these data the best fit to the data in each case was obtained by use of the NUMEG model fitting procedure.

For the pre-replant condition the dataset consisted of 59 measurements (p, see section 3.1). The standard deviation (s) associated with each measurement is an estimated value intended to take account of spatial variations (including boiler-to-boiler), temporal variations and instrument error. The HP and IP turbine swallowing capacities (y) were allowed to vary within sensible limits. In the achieved fit the number of relative errors greater than 1.0 was 14 and the number greater than 2.0 was 5; if the errors were normally distributed the numbers expected would be about 19 and 3 respectively. The results therefore suggest that an adequate fit had been achieved.

For the post-replant condition the dataset consisted of 65 measurements. The HP and IP turbine swallowing capacities were again allowed to vary in order to enable the effective values for the replanted turbines to be derived. In the achieved fit the number of relative errors greater than 1.0 was 21 and the number greater than 2.0 was 6, indicating that an adequate fit had been achieved.

On the basis of the pre-replant and post-replant fits, NUMEG optimizer runs were carried out to assess the improvements in optimum net electrical output due to the replanted turbines. This was done on a consistent basis; i.e. in both cases the assessment assumed plant constraints and conditions (reactor parameters, cooling water temperature, carbon deposition levels) to correspond to current conditions. That is, the current capability of the plant was compared with the capability which would have applied at the same date if the original turbines had been retained.

The gain in net electrical output was found to be in reasonably close agreement with that derived in the design assessment. An important factor is that the swallowing capacity of the new HP turbine is considerably higher than that of the original turbine; before the replant, the output was limited by the constraint on maximum feed pump discharge pressure, whereas after the replant this ceased to be an active constraint.

5. ACKNOWLEDGEMENT

This paper is published by permission of British Energy plc. Copyright of this paper has been assigned to the Institution of Mechanical Engineers.

6. REFERENCES

1. P. E. Berry, K. E. Crisp and C. R. Gane, NUMEG: A Steady-State Whole Plant Model for Nuclear Power Stations, 1st Int. Conference on Advanced Computational Methods in Heat Transfer, Vol.3, Portsmouth, UK, 1990.

2. A. Anderson, C. R. Gane and B. Scruton, NUMEG Suite of Whole Plant Models for Performance Optimisation and Condition Monitoring, paper presented at EPRI Nuclear Plant Performance Improvement Seminar, Ashville, North Carolina, USA, September 1996.

3. P. Lightfoot, R. T. Deam, C. H. Green and J. Rea, NUMEL: A Computer-Aided Design Suite for the Assessment of the Steady-State, Static/Dynamic Stability and Transient Responses of Once-Through Steam Generators, Paper 34, BNES Conference on Boiler Dynamics and Control in Nuclear Power Stations, Bournemouth, UK, 1979.

4. K. E. Crisp and C. R. Gane, A Scaled Lagrangian Method for Performing a Least Squares Fit of a Model to Plant Data, 6th Int. Conference on Numerical Methods for Thermal Problems, Swansea, UK, 1989.

5. K. H. Ardron, J. N. J. Caplin, C. R. Gane and M. Donne, Use of On-Line Total Plant Model for Plant Performance Monitoring, paper presented at EPRI Nuclear Plant Performance Improvement Seminar, Charleston, South Carolina, USA, August 1994.

6. I. C. Parmee, C. R. Gane, M. Donne and K. Chen, Genetic Strategies for the Design and Optimal Operation of Thermal Systems, paper presented at EUFIT '96 4th European Congress on Intelligent Techniques and Soft Computing, Aachen, Germany, September 1996.

S593/003/98

PROATES™ – whole power plant modelling software package for optimizing performance

C H GREEN, A B READY, P E CHEW, and **K HARTWELL**
Power Technology Centre, PowerGen plc, Nottingham, UK

SYNOPSIS

PROATES is the PowerGen proprietary package for modelling the thermal performance of power plant. It represents twenty years of development and application activity on a wide variety of plant types spanning traditional boiler/steam turbine plant through to more advanced cycles such as those associated with coal gasification and fuel cell systems. This paper describes some recent example applications and some of the areas of current development activity.

1 INTRODUCTION

The PROATES package has been developed primarily for predictive analysis of power plant performance for any type of power plant. To carry this out effectively requires an adequate representation of the fundamental physical processes occurring within a wide variety of plant components. This is available within PROATES via a flexible modular structure allowing representation of such plant as traditional coal-fired, IGCC, FBC, CCGT, Fuel-cell, HAT cycle etc. The physics is extensively validated and the PC based modelling package itself is fast, robust and flexible in use. There is also a significant experience base of application knowledge supporting the software. The following sections describe some recent applications of PROATES and also give an overview of some aspects of the current development programme.

2 MULTI-FUEL FIRING IN A CONVENTIONAL BOILER INVOLVING SIMULTANEOUS COMBUSTION OF INDUSTRIAL OFF-GASES WITH TRADITIONAL FUELS

This application concerns the boilers at an industrial plant providing process steam and a proportion of the electrical requirements for the site. The boilers are capable of firing with blast furnace gas (BFG), natural gas (NG) and heavy fuel oil (HFO) or a combination of these fuels. The boilers were operating with a final superheat temperature of approximately 518°C compared with the design of 538°C and, for historical reasons, with the final feedheaters out of service. Early in the life of the plant the final superheat steam operating temperature had been reduced because of final superheater tube failures. Since the replacement of the final superheater and operating at reduced final steam temperatures there have been no tube failures. The reduced steam temperatures mean that operating costs are increased due to increased requirements for fuel enrichment (and reduced BFG use). An efficiency loss is also incurred due to operation without the final feedheater in service and this also increases the demand for BFG enrichment using NG. The purpose of the study was to understand how to optimise plant performance in achieving the required electrical output within plant constraints while at the same time minimising the use of bought in natural gas/heavy fuel oil since supplementary firing with natural gas and/or HFO displaces BFG use and increases operational costs (note BFG is virtually free). To assist in the understanding of options for minimising the use of NG enrichment whilst maintaining maximum power output from the turbo-generator a PROATES™ boiler model was constructed and was used to investigate various options for improving performance.

2.1 Boiler Modelling and Plant Constraints

A view through one of the boilers is shown in Figure 1. A detailed PROATES™ boiler model was constructed which also includes sufficient representation of the back pressure turbo-generator to accurately represent all of the boiler interactions with the plant.

Some particular features of the modelling are:

- A furnace representation made up from several (9) furnace section modules allowing a detailed representation of the multi-fuel firing and air injection points.
- Cavity and tube bank modules representing the exchange between the gas and boiler fluids.
- The furnace, cavity and bank modules all contain detailed formulations describing radiation and convection heat exchange within and between modules.
- Three other modules, an airheater, a turbine and a feedheater complete the representation.

The following assumptions were made and constraints imposed in the analysis:

- Each Unit is run to produce 10MWe of load out of the back pressure turbine which is fed with steam from the boiler at 133 bar (absolute).
- It is observed that increasing use of BFG increases the maximum tube metal temperatures in the 4th stage S/H. This is assumed to arise due to the increase in gas flow rates with increased use of BFG which prevents across boiler gas temperature maldistributions (arising during combustion) from being attenuated by the effects of heat transfer as much as they would have been if the gas flows were lower. Analysis of available data indicated a linear relationship between (peak metal temperature - mean steam temperature) and combustion gas flow and this relationship was used within the boiler model to estimate peak 4th stage metal temperatures. The maximum allowable 4th stage S/H metal temperature is 620°C.
- The maximum superheat spray flow is 9 kg/s.
- The reheat spray flow is not limited but is typically 1.25 kg/s.
- Economiser water outlet temperature is limited to 319°C to ensure an adequate degree of feedwater subcooling to the drum. This value has been taken from the supplied operating data and is equivalent to a subcooled margin of 14°C.

- The minimum airheater metal temperature is assumed to be equivalent to the average of the air inlet and gas outlet temperature. Modifications to the airheater on one of the units enables lower average air in/gas out temperatures to be tolerated without causing corrosion. The minimum average air in/gas out temperatures allowable are 134°C and 125°C respectively for the case where the airheater has not been modified and for the case where the airheater has been modified.

Plant data were used to establish the boiler model characteristics.

2.2 Conclusions

A wide range of options were studied and the main conclusions were:

- The natural gas enrichment required to achieve the required output with average airheater air inlet/gas outlet temperatures at 134°C, final superheater steam temperatures at 518°C, no HFO firing, and with no feedheater in service is estimated to be 13.9% by weight. In this case and all other cases considered, superheater spray flow is on its constraint limit of 9 kg/s to allow minimum enrichment.

- Reduction of the average airheater air inlet/gas outlet temperature to 125°C improves efficiency by 0.9% points but has only a marginal effect on the natural gas enrichment requirements.

- Increasing the final superheater outlet steam temperature from 518°C to 538°C reduces the natural gas enrichment requirements to 6.9%, but also increases the 4th stage peak metal temperatures. The increased peak metal temperatures are within the current design limits of 620°C.

- Reinstatement of the final feedheaters combined with increasing the final superheater steam temperature to 538°C reduces the natural gas enrichment requirements to 3.1%.

- Firing with HFO reduces enrichment requirements and allows approximately 7 kg/s extra BFG to be burnt. Thus it may be economically attractive to burn HFO rather than NG as an enrichment fuel. Certainly, if HFO costs are less than or equal to natural gas costs then burning HFO is of benefit.

- Reductions in the 4th stage superheater surface area of up to 20% could reduce the need for natural gas enrichment from 3.1% to 0.9% (no HFO firing).

- Temperature distributions in the 4th stage superheater could be evened out by reducing the length of tubes in elements which consistently show the highest temperature if this is shown to be necessary, when the boiler is operated at its design steam temperatures, with the feedheaters in service.

Based on the analysis, recommendations were made covering the range of possible plant development options and these are being evaluated.

3 ANALYSIS OF THE PERFORMANCE IMPACT OF LOW NO_X RETROFIT OPTIONS ON A CONVENTIONAL PF BOILER

The purpose of the study was to investigate the effects on overall boiler performance and operation of proposed NO_x reduction schemes for a 500MWe front wall pf fired boiler with a single rear pass. The effect of modifying the furnace to introduce reburning and overfire air was considered for a number of different cases in which various fuel types and furnace configurations were imposed.

3.1 Boiler Modelling and Assumptions

A schematic of the boiler is shown in Figure 2. A PROATES™ model was built with the following specific features:

- The furnace was constructed from 11 furnace section modules thus allowing for the inclusion of the normal burner belt fuel and air input, the injection of reburn fuel using rear pass gas or steam as the carrier, the addition of overfire air and the modelling of the furnace shoulders, nose and exit sections.

- The main fuel considered was either coal or gas and the reburn fuel was chosen from either coal, gas, Heavy Fuel Oil (HFO) or Orimulsion.

- Cavity and tube bank modules represent the exchange between the gas and boiler fluids.

- The furnace, cavity and bank modules all contain detailed formulation describing radiation and convection heat exchange within and between modules.

- Reheat temperature control is primarily achieved via a shell and tube heat exchanger which exchanges heat from the reheat steam to the economiser outlet water with spray attemperation as a secondary means of control.

- Three other modules, an airheater, an HP turbine and a feedheater together with a Willans line correlation for generated output complete the representation.

Configurations considered were:

- Firstly to use the present fourth burner row for reburn fuel injection, the position of the overfire air ports being dependent on the reburn fuel type. Reburn air flow was controlled to produce a reburn zone stoichiometry of 0.9. Reburn stoichiometry is defined as the ratio of the actual total oxygen supplied in the main burner and reburn zones to the stoichiometric air required to completely burn all the fuel supplied in these two zones. Airheater gas inlet O_2 levels were set to the typical levels of 3.2% and 2.1% v/v (wet) for coal main firing and gas main firing respectively. Flue gas recirculation from the economiser outlet was fixed at 29.7 Kg/s.

- Secondly to employ all four burner rows for main firing, with reburn injection points at above the burner belt. Overfire air ports were positioned at various heights above the reburn fuel injection point depending on whether the reburn fuel was Orimulsion, HFO, coal or gas. Since the furnace residence time for injected coal particles at the reburn height is then small, it was proposed to consider use of a microniser to pulverise and transport the reburn coal. The microniser employs steam to pulverise and transport the coal to the burners. For this reason a steam injection point (steam flow 48 te/hr, 9 barg and 175°C) was included in the model at the reburn fuel injection level.

3.2 Conclusions

The base case against which other cases have been compared represents the current operating condition with coal firing in 8 burners on each of rows 1 and 3 and in 4 burners on rows 2 and 4 where row 1 is the bottom row and row 4 the top row. There is no reburn fuel or overfire air and the excess oxygen at the airheater inlet is set at 3.2% by volume in dry gas. For this and all other cases the generated load is fixed at 520MWe.

A significant number of fuel and other parametric variations were analysed and the following main conclusions drawn:

- Generated efficiencies vary from -0.83% to +0.45% when compared with the base coal case (40.35% based on generated output and NCV). The greatest contribution to reduction in efficiency results from increased reheat spray flows. (Stack temperatures being acceptable). A practical way to reduce the reheat spray flows and hence increase efficiency would be to lower the furnace firing level – by biasing more fuel to the lower burners.

- Comparing the second configuration with the first, reheat spray flow requirements are higher for the second configuration. This is due to a higher mean furnace firing level, which has a detrimental effect on efficiency.

- The high overfire air port elevations, proposed as part of the second configuration, may adversely affect the risk of local furnace tube failure, due to critical heat flux limitations being exceeded.

- The effect of the microniser is a notable increase in gas mass flow and reduction in furnace exit temperature. As a result, cases which include a microniser have the highest spray demand and hence lowest generated efficiency of all the options considered.

- For coal main firing, using HFO as the reburn fuel results in a higher generated efficiency than using gas, coal or Orimulsion. All reburn cases tend to reduce the furnace gas exit temperature and increase the gas mass flow thus reducing the heat transferred in the furnace and increasing the heat transferred in the reheater and convective zones thus increasing the need for reheat sprays which causes a loss of efficiency. HFO reburn fuel has the least effect in this regard compared with the other reburn fuels.

- Compared with coal, if gas is used as the main firing fuel a higher NCV based generated efficiency is achieved. However when comparison are made on GCV based generated efficiency then coal firing will give a higher efficiency compared with gas.

- Superheater steam temperature control is possible within normal control limits. The cooling effect available from the non-contact desuperheater is insufficient and reheat spray control is required. However, the need for reheat spray may be reduced or eliminated by altering the main burner firing pattern to lower the main burner heat release. Some superheat control is required in all cases, the lowest spray flow demand being for the gas burning cases.

Overall the PROATES™ analysis has allowed consideration of boiler low NO_x retrofit options to be carried out with confidence in relation to the impact on plant constraints and generated efficiency.

4 REPOWERING, CO-FIRING AND COAL QUALITY APPLICATIONS

Some other recent PROATES application examples which cannot be covered here due to lack of space are outlined below.

4.1 Parallel GT Repowering of a Large Orimulsion-Fired Plant

This application describes the review of a generating unit firing Orimulsion with the aim of finding options for increasing its output and efficiency to allow extension of economic life. The 500MWe unit was originally designed for oil firing, and had been modified to burn Orimulsion some years previously. Under Orimulsion firing, the Unit had typically been generating 450MWe.

In this parallel repowering approach a GT and HRSG are installed alongside the existing unit, and the steam from the HRSG is fed into the existing steam turbine train in parallel with the steam from the old boiler. PROATES™ was used to predict the performance of the Orimulsion-fired unit when parallel repowered with the latest class of GT and investigate the possibility of using an adjacent mothballed boiler as the HRSG in this application.

4.2 Co-firing of Gas and Coal in a Boiler Designed for Coal

This application used PROATES™ as part of the analysis examining an option for gas co-firing in a 500MWe boiler originally designed for pulverised coal. The boiler was front wall fired with a single rear gas pass. The purpose of the analysis was to understand the impact on efficiency, plant controllability and local plant conditions. One issue which did arise was the ability to maintain reheat temperatures under clean surface conditions.

4.3 Impact of Coal Quality on the Output of a Brown Coal Fired Plant

This example of the application of PROATES™ arises from a due diligence review that was being carried out on a particular plant. The plant consisted of two units, of nominal output 350MWe each, burning a very high moisture, low heating value brown coal. It was known that there were problems with the generated output from the plant in relation to coal quality, due to limitations in the milling capacity. The changes in output came from changes in coal moisture content. In order to investigate this, a PROATES™ scoping model was set up to provide a general representation of these units. This was not an exact representation, as details of the boiler banks, steam temperature control systems etc were not known. However, the model as employed was expected to provide reasonable estimates of sensitivities and is an example of using simplified models for scoping studies. In this case, PROATES™ was used as a tool to confirm the behaviour of the plant in relation to changes in fuel moisture and the result of a planned plant upgrade. This was achieved with a generalised model of the boiler turbo-generator unit and with knowledge only of the main plant parameters (steam conditions etc).

5 ANALYSIS OF PERFORMANCE SENSITIVITIES ON ADVANCED SUPER-CRITICAL PF PLANT

A reference model of an advanced supercritical pulverised fuel (pf) plant has been constructed using the PROATES™ modelling system and this has then been used to evaluate a wide variety of plant performance sensitivities. The work was carried out as part of a programme of evaluation of a wide range of advanced clean coal technologies including other technologies such as IGCC and FBC.

6 FUTURE DEVELOPMENT

PROATES continues to be developed to meet the needs of PowerGen and, increasingly, of the growing external customer base. Two specific development features are:

- Extension of optimising capabilities. Essentially, this exploits mathematical optimisation combined with PROATES' physical modelling capabilities to automate the search for optimum performance options of large complex plant.
- Continuing addition of new modules to represent advanced and novel forms of power generation cycles.

7 ACKNOWLEDGEMENTS

Acknowledgement is made to PowerGen plc, Power Technology Centre for permission to publish this paper.

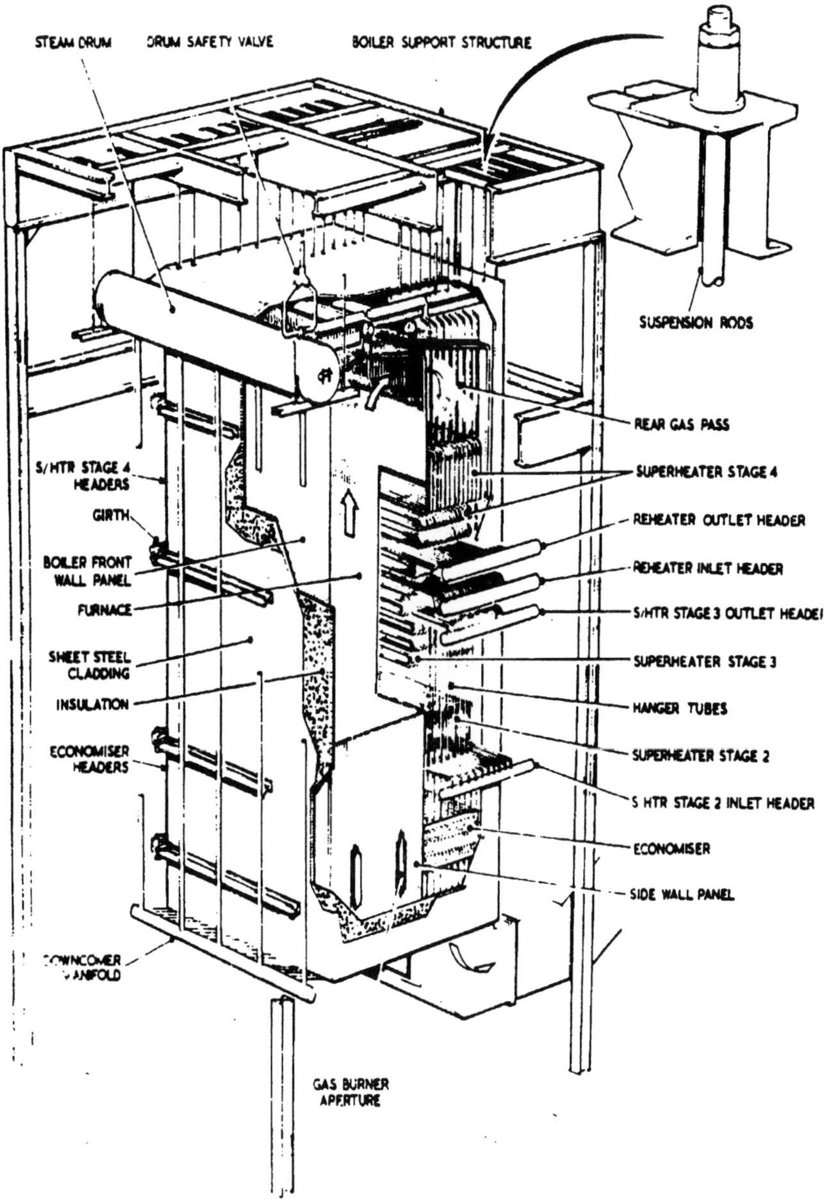

Figure 1: Sectional Drawing of Boiler

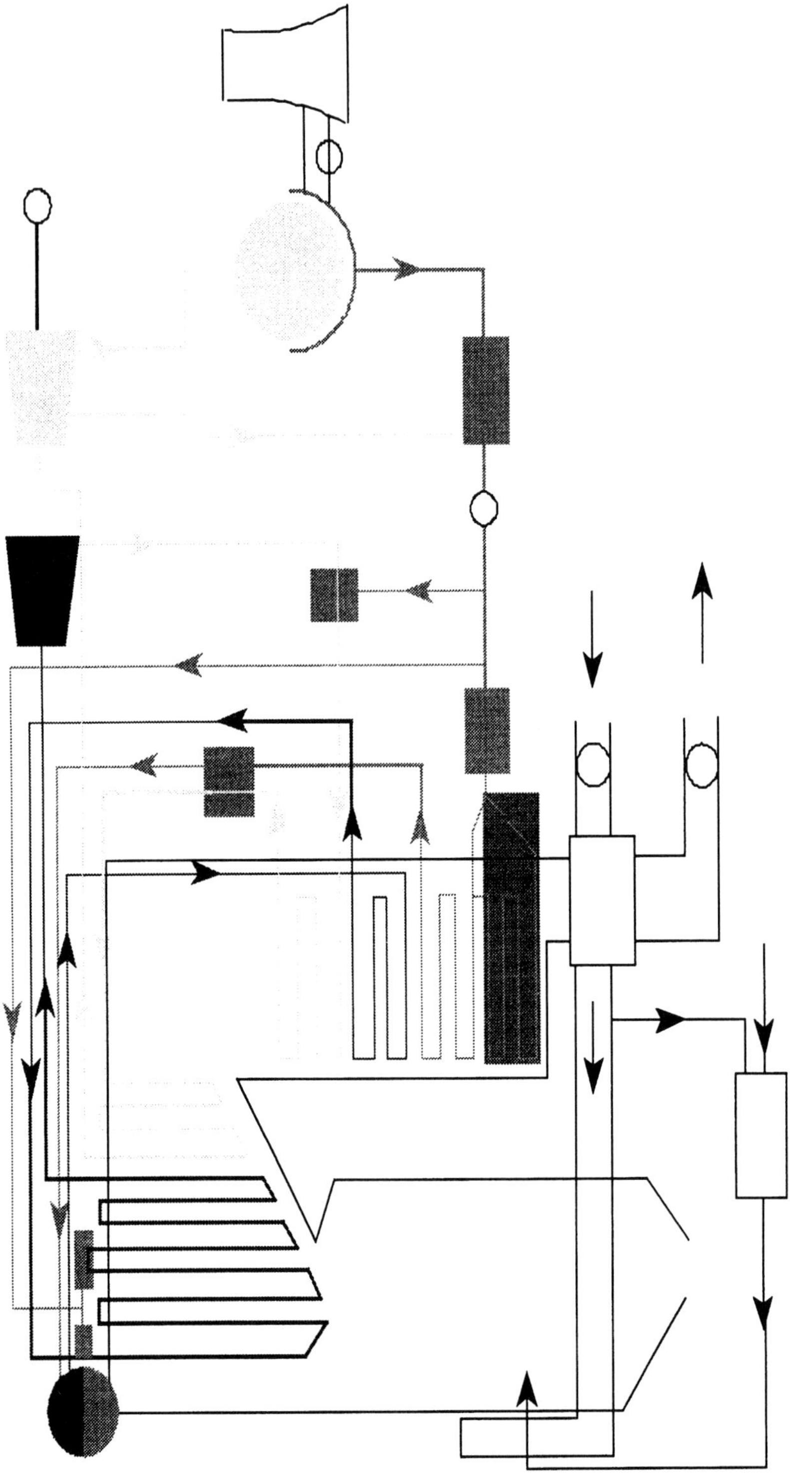

Figure 2: Schematic Representation of the Boiler

S593/004/98

Furnace modelling for anthracite firing

J GILLESPIE
Mitsui Babcock Energy Limited, Renfrew, UK

1. INTRODUCTION

Developing countries in Asia, in particular China, presently have the world's largest growth of installed electrical utility power generation, ordering over 22,000 MW_e in 1996. A substantial proportion of these new power stations will be fuelled by the large reserves of anthracite that exist within the region. Anthracite is very low in volatile matter containing less than 14% by weight measured on a dry, mineral matter free basis. The burning process of anthracites is similar to that of burning carbon. The combustion of low volatile coals does not rely on the volatile component for ignition and combustion stability as does the combustion of higher volatile content fuels such as bituminous coals. The volatiles play a lesser, though important role. It can be assumed that anthracites have two main stages of combustion, viz,

- heating the coal particles to their ignition temperature
- burning the coal

A special type of furnace design is used to burn anthracites. Arch or 'downshot' fired furnaces have been used since the early 1930's to burn low volatile anthracite coals. The term 'downshot' describes the manner in which the fuel is projected into the furnace chamber. This furnace design provides the long residence time within a hot environment that is necessary for their stable, efficient combustion.

Established furnace design methods, whilst based on sound physical principles, have largely relied on empirical rules. This is acceptable where new designs are near or within an 'envelope of experience'. In situations where extrapolation cannot be justified or performance improvements are required, engineers have often turned to physical test work to create new design rules. Unlike bituminous coal firing where stable flames are anchored on individual burners, the anthracite combustion system relies on the interaction between the burners and the furnace as a whole to maintain flame stability. To reproduce these processes physically would require a sizeable, dedicated combustion test facility in which it would be difficult to reproduce full-scale conditions or to determine the influence of small changes to the design changes on, for example near burner flow patterns. An alternative approach has been used involving extensive use of Computational Fluid Dynamics (CFD) to examine alternative furnace and burner configurations and to confirm design features with the following objectives:

- Improved combustion efficiency
- Lower NO_x emissions
- Extended load turndown capability

Designs have been developed which offer operating cost and environmental benefits to operators of this type of plant.

2. DESCRIPTION OF THE FURNACE AND PROCESS

2.1 Basic principle

The original design concept of downshot firing was developed in order to facilitate the combustion of low volatile coals. The design allows the recirculation of hot flue gases to the root of the flame, which is an important feature in controlling flame stability. This is achieved in slot, elliptically shaped, or circular burners firing vertically downwards, relying on furnace shape, high product fineness and refractory coverage to create the conditions necessary to ensure stable, efficient combustion. When firing lower volatile coals, especially anthracites, low NO_x emissions are not so easily achieved. The requirements for stable combustion are at variance with those for low NO_x, in particular, high temperatures are required to rapidly heat the particle, releasing what little volatile matter is available.

To ensure ignition and stable, efficient, clean combustion of the low volatile coal, three criteria must be satisfied, viz;

1. volatile release and initial combustion must be maximised by rapid heating of the air and fuel in the early part of the flame;
2. this must be followed by sufficient residence time in a hot environment so as to complete combustion;
3. an environment conducive to low NO_x emissions must be maintained.

The downshot design relies on radiant heat transfer from the flame and recirculation of a proportion of the hot flue gases leaving the lower furnace to heat and ignite the incoming fuel, and hence stabilise the flame, shown diagrammatically as **Figure 1**. The flame must use the volume of the furnace to provide the relatively long residence time at high temperature to burn out the combustible content of the fuel. Gas temperature and local oxygen concentrations must be controlled to prevent excessive NO_x emission.

2.2 Plant Arrangement

A typical downshot fired furnace arrangement is shown in **Figure 2**. This is a $350MW_e$ boiler with 4 double ended tube mills each serving 8 burner groups. There are 16 individual burners, each comprising 2 slots arranged in 4 pairs arranged for downshot firing on each arch. The distribution of air around the furnace is shown in **Figure 3**. Each burner has an associated cyclone to enrich the primary air-fuel entering the furnace. A controlled proportion of air is removed from the transporting air-fuel stream in the cyclone, this air enters the furnace via slot openings in the arch, away from the primary air/fuel slots, and is known as vent air. Two secondary air slots on either side of the primary air/fuel slot supply the majority of the combustion air. Additional combustion air enters through openings in the lower furnace front and rear walls as tertiary air, and as a curtain of air through slot openings at the junction of the front/rear walls and the arch as wall air.

3. FURNACE MODELLING BACKGROUND

Coal combustor simulation has been of interest since CFD techniques were first developed. It is one of the most complex problems encountered. Behaviour at length scales as small as individual pulverised coal particles and as large as a 50m tall utility furnace must be resolved simultaneously. In the furnace chamber there are closely coupled mechanisms. The aero-dynamics of the burner affect almost all aspects of furnace operation because interactions of the fuel jets with the furnace gases determine how fast the coal mixes with the air and therefore the intensity of heat release and the volume of the primary combustion zone.

The physics of fluid dynamics provide the basis for the simulation. Given adequate definition of the furnace geometry by a sufficiently detailed mesh, and of the incoming flow, the gross flow pattern can be resolved. The K-ε turbulence model was applied in these simulations as it was found that the flow was generally isotropic. The P-1 radiation model was invoked since it allows particle radiation effects to be modelled. Radiation and convection to and from the particles determines the particle heating rate, ignition and hence stability of the flame and it is important that these features are included. The accurate prediction of the point of ignition is crucial to development of a realistic flame shape as it also influences the later coal heat release.

The area of greatest uncertainty is the rate of heat release, this is directly related to the coal properties. In reality there are a number of coal quality related issues which will influence the rate at which the coal will combust. For the purposes of modelling these are greatly simplified and grouped into three processes, viz;

1. devolatilisation;
2. combustion of the volatiles within the gas phase;
3. char combustion.

It could be argued that in hot combustion chambers with sufficient oxygen and intense, efficient mixing, the first two mechanisms are relatively quick (of the order milliseconds). Any errors in the rate of release and combustion of the volatiles will not significantly affect predictions; particularly where the small amounts of volatiles infer a less significant role.

Char combustion is a relatively slow process, the largest diameter particles can take a number of seconds to fully burn out. To represent the char combustion rate a linear series film diffusion-chemical kinetic control scheme is used following Field[1]. Oxygen diffuses from the surrounding flue gas through a boundary layer to the particle surface where heterogenous combustion of the char occurs, represented by a first order Arrhenius equation. The rate at which the heterogenous char reaction proceeds tends to be controlled by diffusion of oxygen to the particle surface in the main furnace volume since gas temperatures are high and the coal is finely ground i.e. smaller diameter particles at the temperature of the surrounding gases. Given the hot, well mixed environment within these downshot fired furnaces and the long residence time of the coal, up to 90% of the combustible content will be released in the lower furnace. It is the remaining 10% or so which must be released in a rapidly cooling atmosphere that is of most concern since chemical kinetic control then becomes dominant and the inaccuracies of the first order Arrhenius equation approach become evident. Activation energies are high and so the reaction is very sensitive to temperature. Care must be taken to ensure a relevant set of constants is used to define the chemical rate control.

Presently a reactivity correction has been applied to estimate the coal constants required to simulate an uncharacterised fuel. This correction converts a known coal reactivity, checked using measured and predicted data for a full scale furnace, to a reactivity which represents the uncharacterised fuel and is based on non-isothermal thermogravimetric analysis (TGA) at low heating rates. Char reactivity was expressed as $\log(A_o)$ at 50% conversion, where A_o is the pre-exponential factor in the Arrhenius equation. This was calculated using a standard value of the activation energy for char oxidation, E, of 130kJmol^{-1} (2),(3),(4), following:

$$(dW/dt)/-W = A_o e^{-E/RT}$$

It is well known that data obtained at low heating rates can be in error when compared to data obtained at higher heating rates closer to those seen in full scale furnaces. While this is a limitation, experience has shown that the relative reactivity measurements are acceptable, however absolute values should not be relied on.

Emissions of oxides of nitrogen produced by pulverised coal combustion consist primarily of NO with smaller amounts of NO_2. These gases are collectively known as NO_x. The NO_x model within FLUENT/UNS was used to investigate NO_x production from anthracite combustion. NO_x can be produced by three mechanisms, viz:

1. Thermal
2. Fuel
3. Prompt

Thermal NO_x is caused by thermal fixation of the atmospheric nitrogen in the combustion air. Its formation is dependent on three factors: oxygen concentration, temperature and the gas residence time at high temperatures. Fuel NO_x is formed by oxidation of the chemically bound nitrogen in the fuel. The proportion of nitrogen converted to NO_x is dependent on the amount and nature of the char and volatile components as well as the combustion environment. Prompt NO_x forms as the result of hydrocarbon radicals combining with atmospheric nitrogen. Normally only relatively small amounts of prompt NO_x are formed in coal flames hence this mechanism was not considered in the present study.

4. DESCRIPTION OF THE MODEL

'Full' furnace combustion modelling is very ambitious, at the limit of the capabilities of the machines and software that we presently use. In the current furnace modelling work the 'as built' arrangement was used as far as possible.

The design flows and conditions were used. Heat transfer at the wall surfaces was more difficult to specify. In a boiler all the principal heat transfer mechanisms take place simultaneously. These comprise radiation between the suspended solid particles and the furnace walls (metal and refractory materials); non-luminous gas radiation from the products of combustion; convection from the gases to the furnace walls and conduction through ash deposits on the walls. Surface characteristics can vary widely. The enclosing furnace walls will include air and fuel inlet arrangements, refractory covered surface, water cooled walls of various constructions. The surfaces may be covered with fluid slag or dry ash of variable thickness. Emissivities of these surfaces will be different and resistance to conductive heat transfer will vary. An overall heat transfer coefficient and emissivity was specified for each wall surface based on analysis of available test data and experience of plant operation

The proportion of volatiles, carbon and inert ash within the remaining dry coal mass was based on the standard proximate analysis for the coals fired. A correction was made to this analysis to account for the additional yield of volatiles due to the high heating rates and temperatures that are experienced by the coal within real furnaces. This correction was defined by work on a High Temperature Wire Mesh apparatus[5], measuring weight loss due to coal devolatilisation at rapid heating conditions (10^4 °C s^{-1} to 1600°C), held for a specified time period before uncontrolled, but rapid cooling.

Particle size distribution has an important influence on the overall performance. At each fuel nozzle a number of representative diameter particles were injected at the velocity of the primary air to simulate the coal size distribution. The coal dispersion was accounted for by use of a stochastic particle tracking algorithm.

5. RESULTS

Flame shape has a major influence on combustion efficiency since the slower burning anthracite fuel requires primarily residence time at high temperature to complete combustion. For downshot firing the resultant flame in double arch units is notionally W-shaped. The thermal environment i.e. time-temperature history of the particles is dependent upon the path the flame takes through the furnace volume and will clearly differ between a 'long' and a 'short' flame. This in turn will effect the volatile release, char combustion and local stoichiometries, all of which will also influence the formation of NO_X. The degree of penetration of the flame into the lower furnace, or the furnace utilisation, depends on the burner design.

5.1 Existing Furnace Design

The existing furnace arrangement was simulated at two different boiler loads. This corresponds to available test data for a 100% and a 60% load condition. Plant test data for air and fuel flows were used. Analysis of coal samples obtained directly from the pulverised fuel supply to the burners provided sufficient information on the coal composition and particle size distribution to complete specification of the model. The test coal composition varied during the tests, a typical analysis is presented in **Table 1**.

	(% w/w) As Received			(% w/w) As Received
Fixed Carbon	62.5		Carbon	63.5
Volatile Matter	8.9		Hydrogen	2.9
Moisture	7.2		Nitrogen	0.8
Ash	21.5		Sulphur	0.5
Gross Calorific Value	24,605	MJ kg^{-1}	Oxygen	3.7

Test Coal Analysis **Table 1**

The furnace flow pattern is illustrated in **Figure 4** which presents a velocity field through a vertical plane at a primary air slot centre-line close to the furnace centreline. The associated gas temperature field is shown in **Figure 5**. Penetration of the flame can be seen extending to the tertiary air level. The flame then turns through 180^0 to exit the lower furnace but also generates and feeds a recirculation zone in the furnace hopper. As the flue gases move past the arch, a proportion of the hot gases are drawn towards the burner slots along the arch and are entrained into the incoming jets serving to heat the air and fuel particles, ignite the fuel and hence stabilise the flame.

A complex pattern is revealed close to the burners, **Figure 6**. The secondary air entrains the recirculated flue gas and primary air-fuel mixture. This emphasises the importance of understanding the influence of burner design features on flow patterns and mixing.

5.2 Comparison against test results

Figure 7 illustrates the measured and predicted performance of the unit in terms of unburned loss (expressed as Carbon in Ash) and NO_x at the furnace exit at two loads. It is clear that there are discrepancies, particularly with the lower load case. However the overall comparison is good. Additional information on ignition (not presented here) also agreed with plant data, indicating the predicted flame ignition level was similar to that measured on site.

It was evident from the site data that there was some burner to burner variation in air and coal flows, this may account for some of the observed discrepancies. There is also some uncertainty as to the wall thermal boundary conditions as the analysis of the measured performance data only provides a gross distribution of heat transfer coefficient and emissivity figures, from experience on site furnace wall surface conditions were variable.

5.3 Influence of Burner Design on Flame Shape

The baseline burner exhibited a relatively long flame with penetration of the fuel and air to the level of the tertiary air injectors. An alternative arrangement was examined with the intention of establishing the effect of shortening the flame length on the furnace performance. **Figures 8** and **9** illustrate the performance of this 'short flame' burner. It is clear from these figures that the alternative burner design arrangement causes the flame to turn earlier using less of the lower furnace, this in turn increases the gas temperature at the firing arch level (exit from the lower furnace).

Figures 10 and **11** present the results of a series of simulations firing a range of coals of volatile matter contents from 2% dry, mineral matter free (dmmf) to 8.2% dmmf, in which different flame lengths were produced by deliberately altering the burner design. It is clear that shortening the flame reduces the heat absorption in the lower furnace and with the poorer utilisation of the furnace causes higher peak gas temperatures leading to increased NO_x levels. It is clear then that short flame length burners generate higher NO_x emissions than long flame burners.

Unburned carbon levels, **Figure 12**, were found to be largely unaffected by gas temperature within the lower furnace. It can be concluded that reductions in NO_x emissions can be achieved by using long flames without adversely affecting combustion efficiency.

6. CONCLUSION

The study has shown that it is possible to predict to an acceptable degree of accuracy the global furnace performance in downshot fired boilers. Further data is required to fully test the model. However based on operational experience it is believed that the CFD model is accurately representing flame shape and by inference the effect of burner design on mixing, heat release and pollutant formation.

This paper has shown that high intensity, rapid mixing burners produce short flames and promote high temperature zones within the lower furnace. Lower intensity controlled mixing burners have longer flames that slow combustion, increase furnace penetration and therefore improve furnace utilisation. This results in lower peak temperatures and reduced NO_x emissions without adversely affecting combustion efficiency or stability.

7. REFERENCES

1. Field, M.A., Combust. Flame, 14:237 (1970)
2. Charpenay, S.,Serio, M.A., and Solomon, P.R., 24th Symp. (Int.) Comb., 1189-1197 (1992)
3. Smith, I.W., Fuel, 57, 409 (1978)
4. Suuberg, E.M., Wojtowicz, M., and Calo, J.M., 22nd Symp. (Int.) Comb., 79-87 (1988)
5. Gibbins, J.R., Man, C.K., Russell, N.V., and Williamson, J.,
   'A Wire Mesh based standard test to characterise coals for PF combustion'
   Imperial College of Science Technology and Medicine
   November 1997

Flame Stabilisation
Figure 1

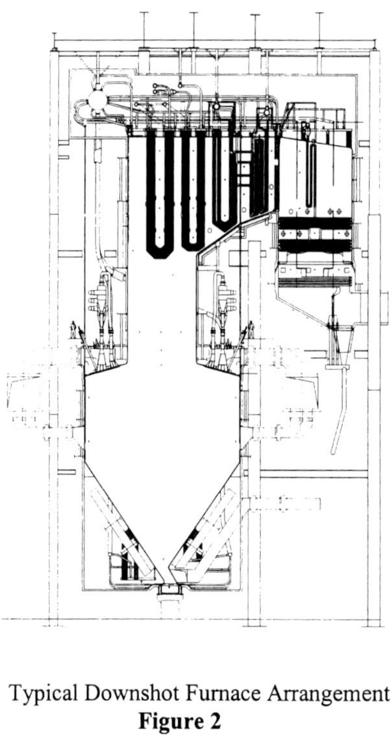

Typical Downshot Furnace Arrangement
Figure 2

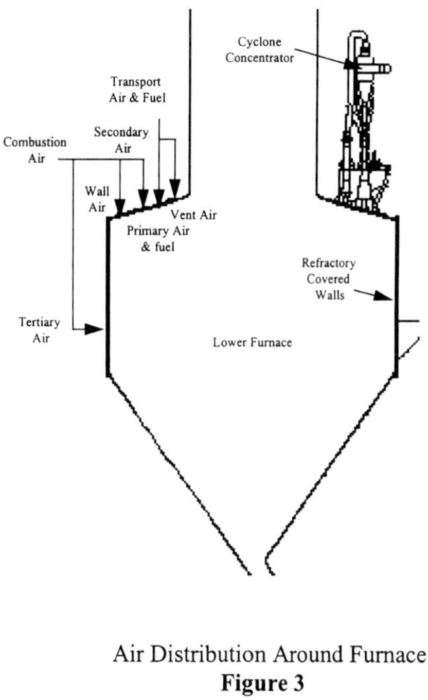

Air Distribution Around Furnace
Figure 3

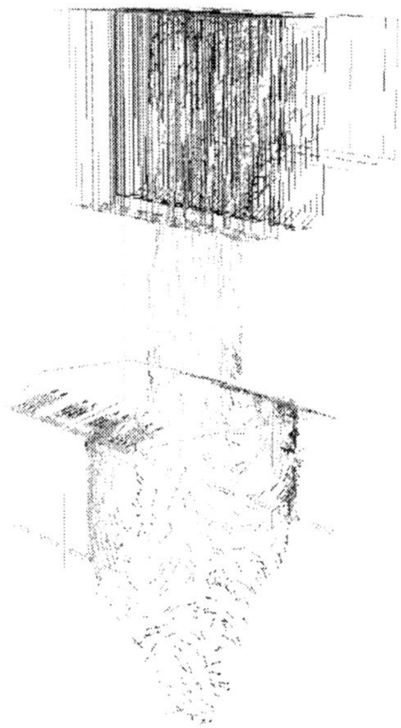

Flow Pattern
Figure 4

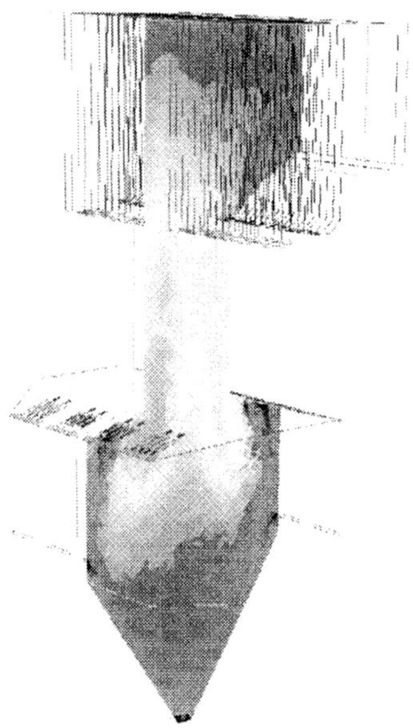

Gas Temperature
Figure 5

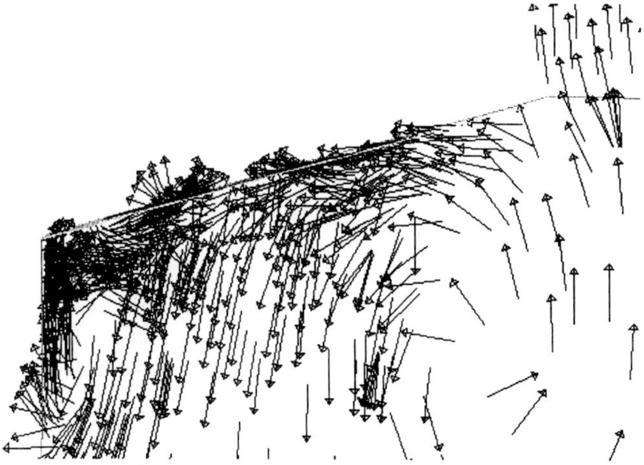

Near Burner Flow Pattern
Figure 6

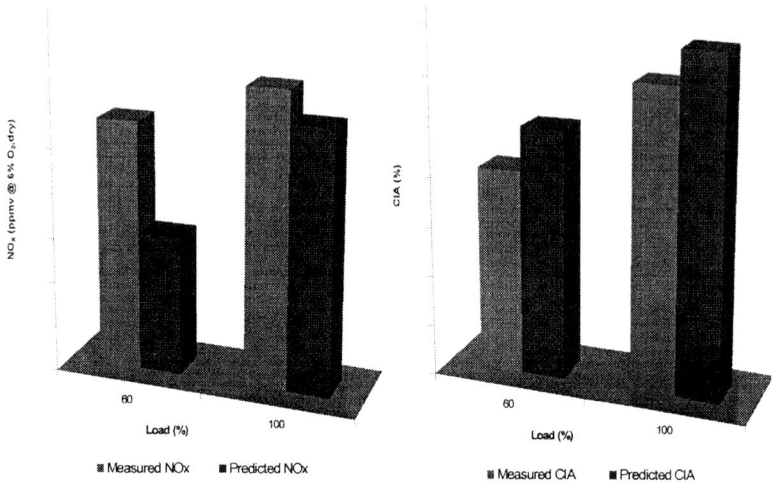

Plant Performance - Measured and Predicted
Figure 7

Flow Pattern
Figure 8

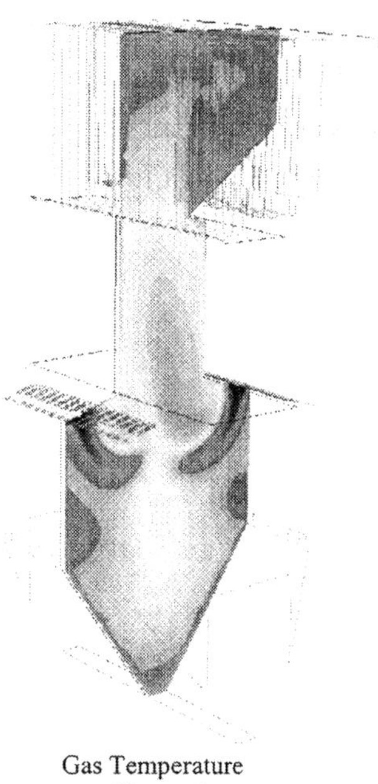

Gas Temperature
Figure 9

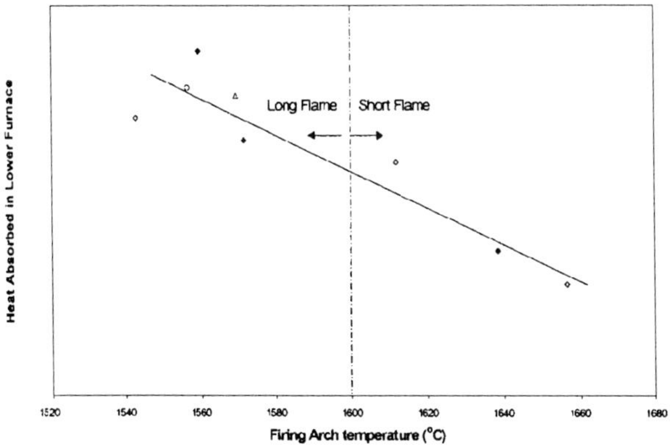

Heat Absorbed in Lower Furnace

Figure 10

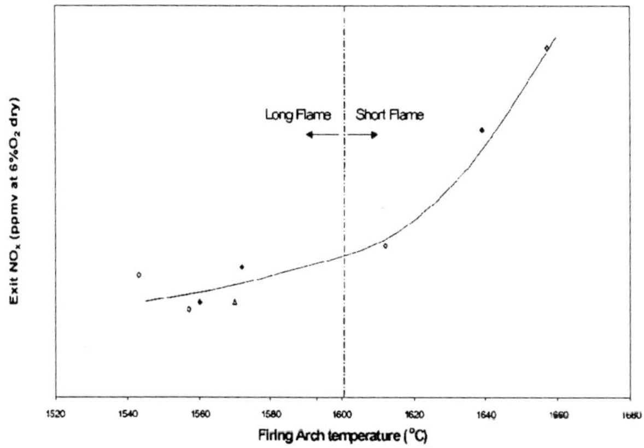

NO_x at Furnace Exit

Figure 11

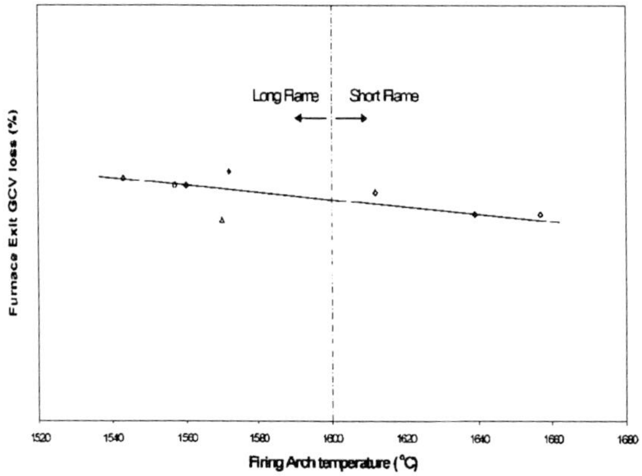

Combustion Efficiency

Figure 12

S593/005/98

Modifications to an electro-hydraulic governor for use with steam bypass system

J P RUSSELL, P H SCOTT, and **R SMITH**
Siemens Power Generation Limited, Newcastle-upon-Tyne, UK

SYNOPSIS

A contract for two 250 MW turbine generators awarded in 1992 included the requirement for an HP and IP/LP steam bypass system. Models of the steam plant and controllers were used in simulations of the sequence and modulating control of the plant to check through the control strategy as influenced by use of the bypass system. In particular, the additional features of the Electro-Hydraulic Governing required by operation with the bypass system were tested by simulation well in advance of plant commissioning, allowing unanticipated problems to be avoided. This paper describes the method of simulation and gives some of the more valuable results.

1. INTRODUCTION

In 1992 NEI Parsons Limited was awarded a Contract by CESC Limited to supply two 250MW turbine-generator units for a new power station in Calcutta, India. The plant was to have fossil fired boilers and the three cylinder turbines were to have an HP bypass system and an IP/LP bypass system. These bypass systems would allow the boiler and turbine to be loaded virtually independently.

Basic control strategies for the HP and IP/LP bypass systems were devised in consultation with the supplier, Sulzer Thermtec Ltd. (now CCI AG) It was then necessary to include some additional control functionality in the Electro-Hydraulic Governor (EHG) to ensure, for example, that steam flows through the turbines during run-up with the bypass system in operation, would be sufficient to prevent damage from occurring. As a means of testing the overall control strategy prior to the building and commissioning of the plant, it was decided that the plant should be mathematically modelled enabling a computer to be used to simulate its operation and that of its associated controllers.

The main areas of concern to be addressed by the modelling and simulation exercise were:

> To ensure that the bypasses operate correctly and stably during run to speed and loading of the turbine.

To ensure that the steam flow through the HP turbine during turbine run to speed is sufficient, when in bypass mode of operation, to prevent churning from occurring in the HP turbine.

To ensure that a smooth transition from bypass to non-bypass mode and vice-versa occurs.

To ensure that the LP bypass maintains a satisfactory pressure ratio across the HP turbine when operating in bypass mode.

To ensure correct operation of the bypasses after the occurrence of a turbine trip.

2 METHOD OF MODELLING AND SIMULATION

The MathWorks Matlab/Simulink software package was used to construct the plant model. This software package has been used by Parsons/Siemens for several years. It is extremely flexible and allows complex models of dynamic systems to be constructed from basic building blocks such as time lags and gains in a hierarchical manner. The model is constructed using Simulink graphical elements, consisting of a series of sub-systems linked by lines which indicate plant and control signal flows between the sub-systems. The graphical model is supported by files which, for example, automate the procedure for setting data, calculate the initial conditions, run the simulation with an appropriate differential equation solver, and present results in the form of graph plots.

2.1 Model Description

Figure 1 shows a schematic of the items of plant modelled. The model represents a conventional fossil fired boiler producing steam at a controlled pressure to the HP turbine via four governor valves. Steam from the HP turbine exhaust is reheated by the boiler and then fed to the IP turbine via two intercept valves. In order to simplify the model, the IP and LP turbines have been modelled as being effectively one turbine. The LP exhaust pressure is assumed to be zero, as a condenser is not modelled. A simple generator model gives the electrical power produced.

Two bypass systems allow steam to bypass the turbines. For the purposes of the model each bypass consists of a control valve and a pressure drop due to pipework and any pressure reduction/de-superheating valves. The HP bypass allows main steam to be bypassed from upstream of the governor valves and Emergency Stop Valves (ESV) to the reheater inlet. The LP bypass allows steam to be bypassed from the reheater outlet to the condenser.

An Electro-Hydraulic Governor (EHG) controls the turbine governor and intercept valves and an HP and an LP bypass controller control the HP and LP bypass valves respectively.

The top level graphic of the model is shown in Figure 2. It consists of sub-systems of the major items of plant. The top level sub-systems are the Turbine Island , the Boiler , the Controllers and the Sequencer . Each sub-system consists of sub-systems itself, and each of these yet further sub-systems. An example of the Boiler sub-system is shown in Figure 3. The Sequencer is essentially a file of C-code which determines the high level sequence control, defining the setpoints, based on the state of plant and the mode of plant operation to be simulated.

2.2 Model Limitations

The model is not suitable for determining the thermodynamic efficiency of the plant or for determining exact data in the steady state. Temperatures are not modelled at all. Therefore it is not possible to use the model for evaluating turbine thermal stress, for example. Also, it is not suitable for simulating power system transients. The model is solely for the purpose of simulating the dynamics of the system in time scales corresponding to the dynamics of the boiler, the steam turbines and the bypass systems.

3. BYPASS AND EHG CONTROL STRATEGIES

The HP bypass and IP/LP bypass controllers operate largely independently of each other, and are independent of any control loops in the EHG. During bypass operation there is a need for some special control functions in the EHG, and in the boiler control systems.

3.1 Boiler Control

The mode of control of the boiler depends on whether or not bypass mode of control is engaged. For non-bypass operation the boiler controller normally maintains the boiler pressure at a setpoint by controlling the firing rate. However when the bypasses are operating, the boiler firing rate is normally controlled to a load setpoint and the bypasses are responsible for controlling the boiler pressure.

3.2 HP Bypass Controller

The purpose of the HP bypass is to pass excess steam produced by the boiler which is not required by the HP turbine. When bypass control is selected by the plant Operator, the HP bypass operates so as to maintain the main steam (boiler) pressure to a setpoint. This will normally occur during turbine start up or shutdown.

During turbine start up, the HP bypass will initially control the boiler pressure to a setpoint. The boiler firing rate will be determined by the boiler load setpoint and the turbine will be loaded according to the rate selected in the EHG. As the turbine load increases towards the boiler load setpoint the HP bypass valve will begin to close down. When the HP bypass closes an offset is added to the HP bypass pressure setpoint, so that the bypass valve remains closed under normal operation, and boiler control is switched from load control to pressure regulation.

If the turbine is tripped or if the boiler is in load control and the turbine is unloaded or shutdown, the main steam pressure will rise. When the pressure becomes greater than the HP bypass setpoint plus its offset the bypass valve will open again. Once the valve has opened, the pressure setpoint offset will become zero and the bypass will control the main steam pressure again.

3.3 LP Bypass Controller

The purpose of the LP bypass is to pass excess steam from the reheater when too much steam is available for the IP and LP turbines. It does this to prevent the reheat pressure from becoming too high, leading to a small steam flow through the HP turbine. A small HP turbine steam flow would result in churning within the HP turbine causing overheating of the blades near its exhaust.

The LP bypass limits the pressure ratio across the HP turbine by preventing the outlet reheat pressure from exceeding a pressure limit setpoint. This limit setpoint is determined as a function of the HP turbine first stage pressure. When the HP bypass is operating, normally during turbine start up or shut down, it will probably be necessary for the LP bypass to operate in order to limit the reheat pressure. During normal turbine operation, with the HP bypass valve closed, the reheat pressure will be below the LP bypass controller setpoint and the LP bypass valves will also be closed.

3.4 Electro Hydraulic Governor

The EHG incorporates four additional functions over and above those which are present in a typical EHG for a unit without a steam bypass system. The special functions are a speed dependent function for maintaining a minimum steam flow through the HP turbine, bypass compensation for the governor power demand, reheat outlet pressure compensation for the intercept power demand and parallel intercept governing mode. Figure 4 shows a simplified block diagram of a typical EHG for a unit without a steam bypass system, and Figure 5 that for the EHGs for the CESC units.

The minimum steam flow function operates only when the machine is in bypass mode, i.e. when the HP bypass valve is open. When operating, the function generates a minimum steam flow demand signal (refer to Figure 5) which is dependent upon the value of the turbine speed setpoint and the reheat outlet pressure. This signal corresponds to the governor valve lift that is required to give sufficient steam flow, at the current reheat pressure and speed, to prevent churning and overheating of the HP turbine blades.

Bypassing steam through the HP bypass may cause the reheat pressure to be higher than normal, causing the steam flow through the HP turbine to be reduced. In order to compensate for the reduction in power caused by the elevated pressure, a Bypass Compensation term is added to the power demand (3) signal. The value of the added term is calculated from the measured reheat pressure and the power demand (3) signals.

The Reheat Pressure Compensation term compensates the intercept power demand for the effect of reheat pressure. Although it is active on the CESC units in both bypass and non-bypass modes, it is not normally present in governors for non-bypass machines.

During bypass mode, i.e. when the HP bypass valve is open, the intercept valves operate in parallel governing mode. Parallel intercept governing mode means that the intercept valve power demands are made equal to those for the governor valves, power demand (4). When bypass mode is disengaged intercept governing action reverts to normal offset governing, whereby the intercept valve demands follow the governor valve demands with a (typically) 50% offset such that when the governor power demand is 25% the intercept valves are fully open. Transfer from parallel to normal offset governing occurs via a ramp function. The same ramp function is used to engage/disengage Bypass Compensation. When in bypass mode it is necessary to use parallel intercept governing because more steam is available for use by the IP and LP turbines than has passed through the HP turbine. It is therefore necessary to limit the steam flow through the IP turbine by use of the intercept valves.

4. SIMULATED SEQUENCES

Over the period during which the model was constructed many simulations were done at the various stages of completion of the model in order to test the model and to test the control strategy for possible problem areas. A series of simulation sequences were defined using C-code. These sequences define how the plant is to be operated during the simulation, i.e. what events are to occur and when they are to occur. Since it is not possible to show the results of all of the simulations performed in this paper, only results from a selection of simulations have been included for illustration. A list of defined sequences for which simulations have been carried out is given in the Appendix.

4.1 Simulation Sequence 1

This simulates the starting up of the turbine without bypass operation. At the start of the simulation the turbine is on barring gear and the boiler is at rated pressure. The speed setpoint for the turbine is raised from -25% to 100% over a 1000 second period, the generator is synchronised and the load setpoint is then increased at a rate of 4 MW per 100 seconds.

4.1.1 Results

The results of the simulation are shown in Plots 1 to 4 (Figure 6). They show the governor valves opening in throttle mode and the intercept valves opening with an offset.

4.1.2 Conclusion

The simulation gave results which were as expected for running up a non-bypass machine, indicating that the model was working satisfactorily.

4.2 Simulation Sequence 3

This simulates starting up the turbine in bypass mode. At the start of the simulation the turbine is on barring gear, the boiler is at rated pressure and at 60% load, the steam produced all being bypassed to the condenser via the HP and LP bypass valves. The speed setpoint for the turbine is raised to rated speed over a 1000 second period, the generator is synchronised and the load setpoint is then increased at a rate of 4 MW per 100 seconds.

4.2.1 Results

Plots 5 to 13 (Figures 7, 8, 9) show the results of the simulation. It can be seen from Plot 7 that the HP bypass begins to close down as governor valves open (Plot 6) and the turbine speed increases (Plot 5). The machine synchronises at 1000 seconds and the loading begins; the governor and intercept valves continue to open, taking more of the steam from the boiler. Plot 8 shows that at 2500 seconds the reheat pressure has reduced to the LP bypass setpoint value and the LP bypass valves begin to close. The HP bypass valve eventually closes at 3200 seconds which results in the transfer of control to non-bypass mode. This causes two things to happen; an offset is applied to the intercept power demand and governor bypass compensation is disengaged.

Plot 11 shows the intercept power demand offset being ramped up to its normal value when bypass mode disengages. This causes the intercept valves to open further. Opening the intercept valves will increase the generated load therefore the MW compensation term becomes more negative, as shown by Plot 12. The effect of exiting bypass mode and therefore adding the normal offset to the intercept power demand is to cause the governor valves to reduce their lift. In addition, the governor bypass compensation term is adding approximately 8% power demand, shown by Plot 10, at the time of disengagement. Disengaging the bypass mode causes this compensation term to be ramped towards zero and the effect of this is to cause the governor valve lifts to reduce. For this particular case, there are therefore two effects which cause a reduction in governor valve lifts when exiting bypass mode. One is due to ramping an offset onto the intercept power demand and the other by ramping off the governor bypass compensation.

The effect of closing the governor valves on exiting bypass mode is to reduce steam flow through the HP turbine, and since the HP bypass is closed, the boiler pressure begins to rise. Disengaging the bypass mode causes the boiler to switch to pressure control mode. The rising boiler pressure therefore causes the boiler firing rate (Plot 13) to be reduced from 60% down to nearly 40% during this time (note that the allowable rate of change of boiler firing was set to 10% per minute in the model, but information has since been received that it is only 3% per minute). Although the boiler firing rate has been reduced, the reduction in steam flow from the boiler is large enough to cause the boiler pressure to rise.

When the HP bypass valve closes, a two percent offset is added to the controller pressure setpoint with the intention of preventing repeated opening and closing of the valve. However, the extent of the reduction in governor valve opening, the high boiler firing rate at turbine start-up and the limited rate at which it can be reduced and the time constants of the boiler (as modelled) results in the sharp rise in boiler pressure to greater than 2% above its setpoint, as shown by Plot 7. In turn, this causes the HP bypass controller to take action by re-opening the HP bypass valve. Re-opening the HP bypass valve constitutes re-entering bypass mode and the processes of adding the intercept offset and ramping off the governor bypass compensation are reversed causing the governor valves to begin to open again. Re-opening the governor valves causes the boiler pressure to fall, thus the HP bypass valve closes down again and bypass mode is disengaged. It can be seen from Plot 13 that when the HP bypass valve opens again, the boiler firing rate increases. This increase in firing rate occurs in response to the increase in steam mass flow from the boiler.

The plots show that after initially exiting the bypass mode it is engaged once more before finally disengaging again. However, the original control strategy was to simply switch off governor bypass compensation and a shorter ramp time was used for applying the intercept offset. This resulted in a repeated process of engaging and disengaging bypass mode for a significant period of time.

4.2.2 Conclusion

This simulation sequence highlighted a potential problem of continuous engagement and disengagement of bypass mode. Modifications to the control strategy have been made as a result to reduce the risk of this happening. The strategy was modified by using the same ramp function, as was used to add the normal intercept power demand offset, to remove the Bypass Compensation term from the governor power demand. As well as implementing this feature, it was recommended that the Bypass Mode Transfer Ramp period be made variable, such that it may be set by the Commissioning Engineer at site and that the range over which it can be set should be greater than the original 40 second value envisaged. Since there is some uncertainty of the boiler dynamics it is not known for sure whether the continuous engagement and disengagement problem would occur at site. This simulation is however the worst case, i.e. starting the turbine at rated boiler pressure with the maximum boiler firing allowed by the bypasses. Simulations at lower initial boiler pressures have shown the problem not to occur, even before the modifications were made to the strategy.

4.3 The Requirement for Minimum Steam Flow

The effect of the Minimum Steam Flow Demand for the HP turbine can be significant during turbine run to speed and initial loading. This can be seen in Plot 9 (Figure 8) where the Minimum Steam Flow Demand exceeds the power demand (power demand (4) in Figure 5) and therefore is controlling the governor valves until approximately 1300 seconds into the simulation, i.e. after synchronising the unit. The influence of the requirement for Minimum Steam Flow may be changed by adjusting the value of a constant, called churn_l, in this sub-system. This constant defines the minimum steam flow requirement at rated speed and rated reheat pressure. For the simulation described in 4.2 above, the value of churn_l was 0.2, which means that at rated speed and rated reheat pressure, the demand produced by the Minimum Steam Flow function will be for 20% of rated steam flow. Further plots for that simulation (Sequence 3) over the first 1000 seconds and plots for churn_l values of 0.3 and 0.1 are discussed in the following sections.

4.3.1 Sequence 3 Results for churn_l = 0.2

This is the value that was used for the simulation referred to in section 4.2. Plot 14 (Figure 9) shows the values of the Minimum Steam Flow Demand and the power demand (4), from which the governor valves are controlled through a Select High function. It can be seen from this plot that the Minimum Steam Flow Demand is controlling the governor valves since the power demand is almost always lower than the minimum steam flow value and therefore has no effect. Plot 19 (Figure 10) shows that the speed of the turbine is however being controlled to its setpoint. This occurs via the droop control term in the power demand, which is only effective on the intercept valves - see Plot 15 of Figure 9.

4.3.2 Sequence 3 Results for churn_l = 0.3

Referring to Plots 17 and 18 (Figures 9/10), it can be seen that by increasing the Minimum Flow Requirement yet further, a point is reached at 800 seconds whereby the Minimum Steam Flow requirement is such that the HP turbine is providing all of the power required to maintain the turbine at the speed setpoint. At this point the power demand has become negative and the intercept valves have closed. At approximately 970 seconds the droop has changed from 8 %, for run to speed, to its synchronised value of 4%. The power demand is then negative enough to cause the governor close down signal (Figure 5) to become effective. This signal is produced from a negative power demand with a large gain. Its purpose is to override the Minimum Flow requirement when the power demand (3) has a large negative value. It causes the governor valves to close at this point in the simulation.

4.3.3 Sequence 3 Results for churn_l = 0.1

It can be seen from Plots 20 and 21 (Figure 10) that with this value of churn_l, the Minimum Steam Flow Demand is never large enough to control the governor valves. Both the governor and intercept valves are therefore controlling the machine speed throughout the simulation. Plot 21 shows the intercept valves to remain open throughout the run to speed and the intercept valve lift is generally greater than that shown by Plot 15, for which churn_l was 0.2.

4.3.4 HP Turbine Minimum Steam Flow

Plot 23 (Figure 11) shows the actual steam flow through the HP turbine compared to that of the calculated minimum flow requirement. The calculated minimum flow requirement is determined using the design value of steam flow through the turbine when at rated speed, unsynchronised, and with rated reheat pressure. This design value is in fact 0.29 pu. Calculation of the required minimum steam flow is then determined by assuming that steam flow generally required is proportional to the turbine speed and the reheat pressure:

RequiredMinSteamFlow = {TurbineSpeed OVER RatedTurbineSpeed}{ReheatPressure OVER RatedReheatPressure}*DesignChurnl*

where DesignChurnl = 0.29.

The above expression was used to generate the minimum steam flow required value given on Plot 23, whereas the minimum steam flow demand is calculated using:

MinSteamFlowDemand = {SpeedSetpoint OVER RatedTurbineSpeed}{ReheatPressure OVER RatedReheatPressure}*Churnl*

where Churnl is a constant which is set such as to make the Minimum Steam Flow Demand equal to, or exceed, the Required Minimum Steam Flow. The value of Churnl used for the model was 0.2. The reason why Churnl is different from DesignChurnl is that the throttle lineariser curve in the EHG does not exactly match the characteristic of the plant, therefore the steam flow through the HP turbine does not match the power demand (7). Plot 23 shows that, for Sequence 3, the Minimum Steam Flow function of the EHG is effective and ensures that the HP steam flow is higher than the required minimum.

4.3.5 Conclusion

The Minimum Steam Flow function of the EHG does maintain a minimum flow as required. However, when the Minimum Steam Flow Demand is effective the governor valve control of machine speed becomes open loop. The speed of the machine is then being controlled via the intercept valves only. This results in the intercept valve opening being smaller than it would otherwise have been and may, if the churn_l value is set too large, cause the intercept valves to close during turbine run to speed. It is therefore essential the value of churn_l should be correctly set at site, otherwise it is possible that overheating could occur in the IP and LP turbines.

4.4 LP Bypass Operation

The purpose of the LP bypass is to prevent, when possible, the reheat pressure exceeding a setpoint which relates to the pressure ratio across the HP turbine. The setpoint is derived by dividing the measured P1 (first stage) pressure by a constant, which is determined from the pressure ratio at rated conditions, and adding an offset to that value. Simulation Sequence 3 was used to obtain results for a turbine start-up and loading with the initial boiler firing at 60% and a modified version of Sequence 3 was used to obtain results for a case with an initial boiler firing of 20%.

4.4.1. Results with Initial Boiler Firing of 60%

Plot 24 (Figure 11) shows the reheat pressure, its setpoint and the lift of both LP bypass valves. The reheat pressure is initially higher than the setpoint because the LP bypass valves can not pass sufficient steam at this pressure to maintain the setpoint pressure. As loading of the turbine begins, the reheat pressure falls and the LP bypass begins to control the reheat pressure to the setpoint at 2500 seconds. After 3500 seconds, the bypasses are closed and the reheat pressure resumes its normal, non-bypass value. Plot 25 shows the pressure ratio as the turbine runs up and the pressure ratio which the LP bypass tries to maintain.

4.4.2 Results with Initial Boiler Firing of 20%

Plot 26 (Figure 11) shows the results of running Sequence 3 with a modified initial boiler firing rate of 20%. Less steam is being produced by the boiler. The effect of this is that the LP bypass is able to reduce the reheat pressure to a greater extent and thus bring the reheat pressure down to the setpoint value earlier, i.e. at 1500 seconds, rather than 2500 seconds. However, the LP bypass still only controls the reheat pressure to its setpoint over a relatively short period of time because exit from the bypass mode is earlier as a result of less excess steam being generated by the boiler.

4.4.2 Conclusion

Since the LP bypass is sized to pass 60% of rated steam flow at rated reheat pressure, it can not always pass sufficient steam to achieve its setpoint at initial turbine start up, if the boiler has a high initial firing rate. While the bypass does not initially achieve its setpoint, it does limit the reheat pressure to below its rated value (for worst case of 60% initial firing).

4.5 Turbine Trip - Sequence 22

Sequence 23 simulates a turbine trip, at rated boiler pressure, 80% load and with bypass operation enabled.

4.5.1 Results

Plots 27 to 32 (Figures 12 and 13) show the results of the simulation. The turbine is tripped 200 seconds into the simulation and, as can be seen from Plots 29 and 30, both the HP and LP bypass valves respond immediately by fully opening. Plot 32 shows that the boiler controller ramps the boiler firing rate back as a result of the increase in boiler pressure. Although the bypasses have operated, the simulation shows that tripping from high loads causes the boiler drum safety valve to open for a short period of time.

4.5.2 Conclusion

The simulation has shown the bypasses to function as expected in response to a turbine trip.

4.6 Turbine Run to Speed Valve Oscillations

During the development of the model it was noted that some simulations produced valve oscillations during turbine run to speed. Closer inspection of graph plots revealed that there were also small amplitude speed oscillations. The problem was traced to an unstable speed control (droop) loop in the EHG. It was found that the gain of the loop was altered by a mismatch between the throttle valve lineariser curve and the actual characteristic of valve, as modelled. This phenomenon reflects what sometimes occurs in practice when the predicted lineariser curve does not match the actual plant.

It should therefore be noted that if an oscillation of this nature does occur at site and it is found that it can be reduced by using a larger value of droop then the throttle lineariser should be checked.

5. OVERALL CONCLUSIONS

The simulation results included in this report and the results of other simulations which are not included have demonstrated that the control strategy satisfies the requirements as defined in section 1.0. The modelling exercise has however resulted in some modifications to the strategy ie the method of transfer from bypass to non-bypass mode and the length of time over which this is done, as described in 4.2.2. In addition, it has raised some concerns which should be noted and checked at the commissioning stage.

When the turbine is run up in bypass mode and the initial rate of boiler firing is high, as may be the case after a turbine trip, there may be a problem of the bypass disengaging and then re-engaging several times before finally disengaging. There is also a possibility that the boiler safety valves could open transiently when bypass mode disengages. Whether or not this actually happens will depend upon the dynamics of the boiler. The effect which causes the oscillation may be reduced by increasing the time over which the transfer from bypass to non-bypass mode occurs. This time was made variable as a recommendation resulting from this modelling exercise.

It is important that the value of the constant used for the determination of the Minimum Steam Flow Demand be correctly set during commissioning. It must be large enough to ensure sufficient HP turbine steam flow, but not too large as this could cause the closure, or virtual closure, of the intercept valves during turbine run up.

Acknowledgements

This paper describes work carried out by the authors for NEI Parsons Limited. Siemens Power Generation Ltd. is grateful to Heaton Power Limited (who retain responsibility for the contract) for permission to publish this paper.

APPENDIX

Summary of Simulation Sequences Available

Run No.	Description of Sequence Simulated	Initial Boiler / pu		Bypass Mode Enabled
		Pressure	Load	
1	Turbine run to speed and loading to rated load.	1.0	0.0	no
2	Turbine run to speed and loading to half rated load.	0.5	0.0	no
3	Turbine run to speed and loading to rated load.	1.0	0.6	yes
4	Turbine run to speed and loading to 0. 4 times rated load.	0.4	0.2	yes
5	Trip the HP & LP bypass valves at 50% rated speed.	1.0	0.6	yes
6	Start synchronised at no load, load to 0.6 rated load and then unload.	1.0	<0.1	no
7	Start synchronised at no load, load turbine to 0.6 time full load, ramp the boiler pressure setpoint to 1.0	0.5	<0.1	no
8	Start synchronised at no load, load to 0.6 full load and then unload. Boiler firing increased to 1.0 and then decreased to match turbine unloading.	1.0	0.6	yes
9	Start synchronised at no load, load to 0.6 full load and then unload. Boiler pressure setpoint ramped to 1.0 after bypass valves closed.	0.4	0.2	yes
10	Trip the generator at full load, allow the boiler load to run back to 20%.	1.0	1.0	yes
11	Trip the generator at 50% load, allow the boiler load to run back to 20%.	1.0	0.5	yes
12	na			
13	na			
14	Start synchronised with turbine at 40% load, ramp boiler load back from 70% to 60% and then trip the bypasses.	1.0	0.7	yes
15	Turbine run to speed and loading to rated load.	0.4	0.0	yes
16	Start unsynchronised at rated speed and ramp speed setpoint back such that it is always 120 rpm less than actual speed.	1.0	0.6	yes
17	Run at steady state in valves wide open condition.	1.0	1.0	no
18	Run turbine to rated speed.	0.46	0.28	yes
19	Run turbine to rated speed and load to full load.	0.46	0.28	yes
20	Start at 80% load, rated pressure then switch to sliding pressure mode, reduce pressure to 80% and then increase back to 100%.	1.0	0.8	no
21	Turbine run to speed and loading. Switch to sliding pressure mode before the bypasses have closed.	1.0	0.6	yes
22	Turbine trip from 80% load.	1.0	0.8	yes

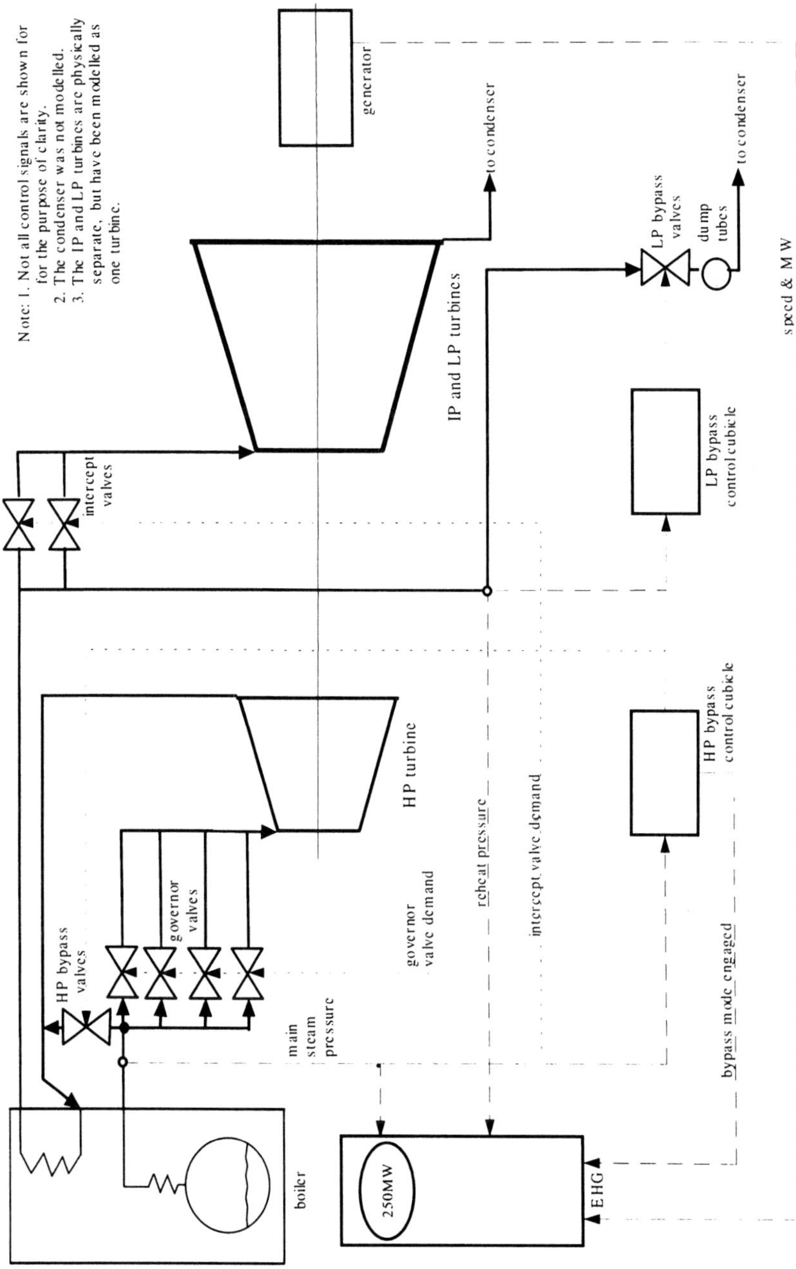

Figure 1 - Simplified Schematic of Plant Modelled

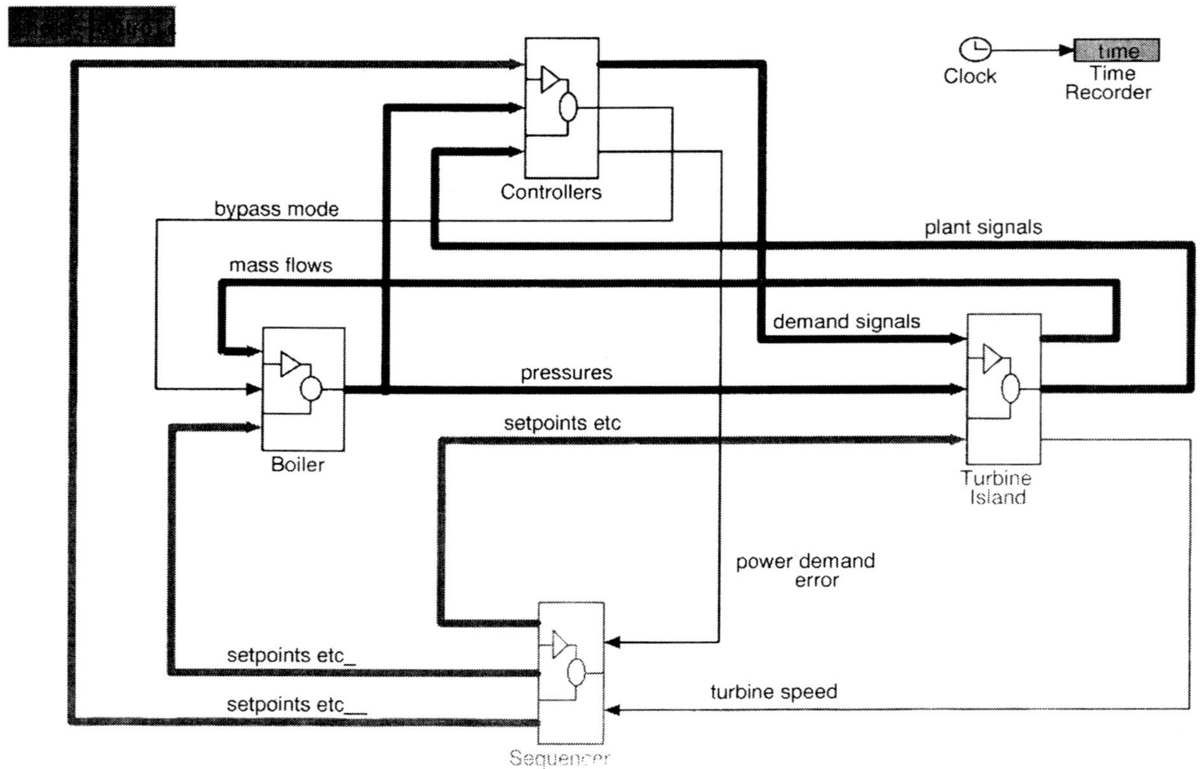

Figure 2 - Top Level Graphic of Simulink Model

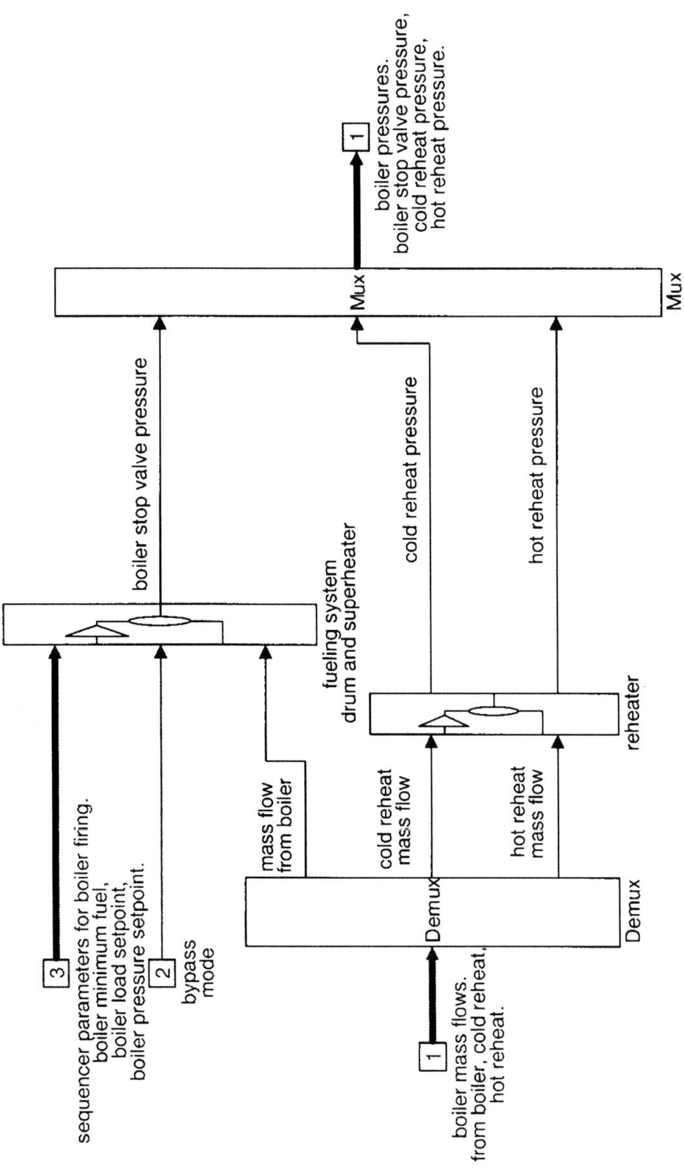

Figure 3 - Boiler Sub-System Graphic

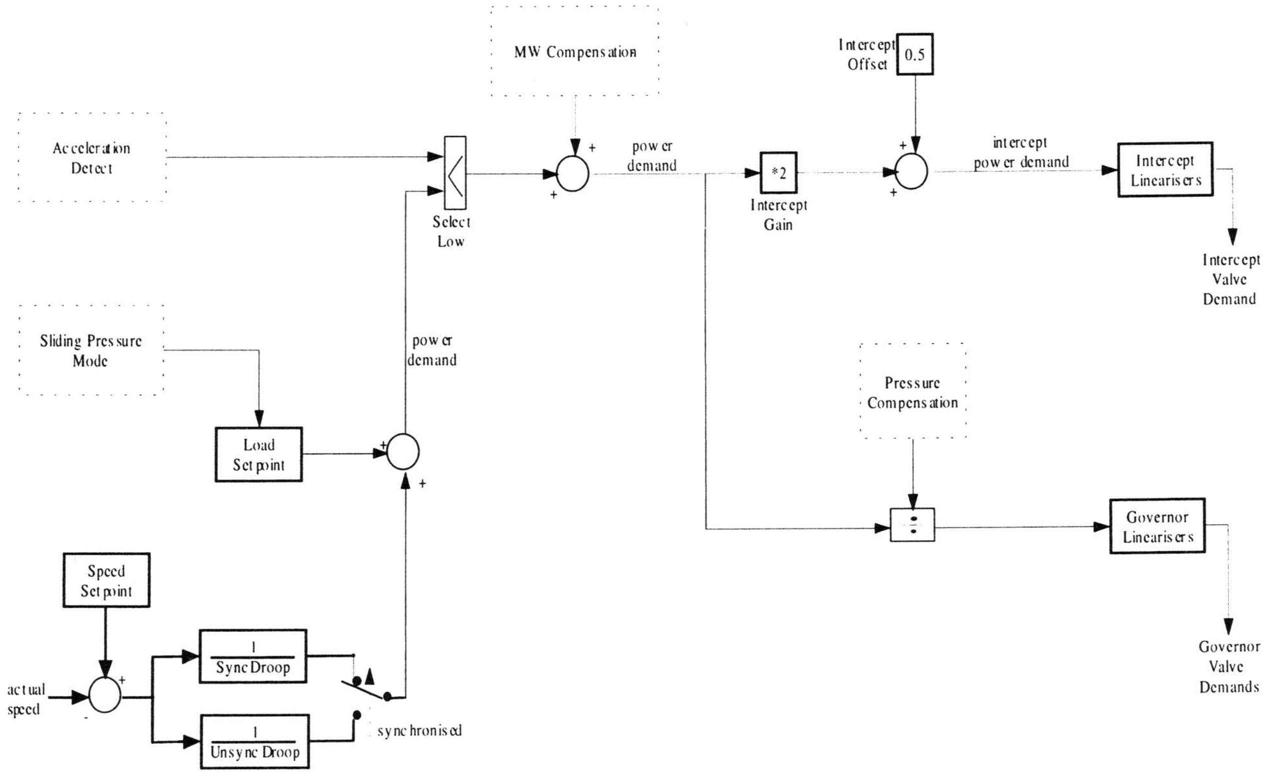

Figure 4 - Typical EHG Control Schematic (non Bypass)

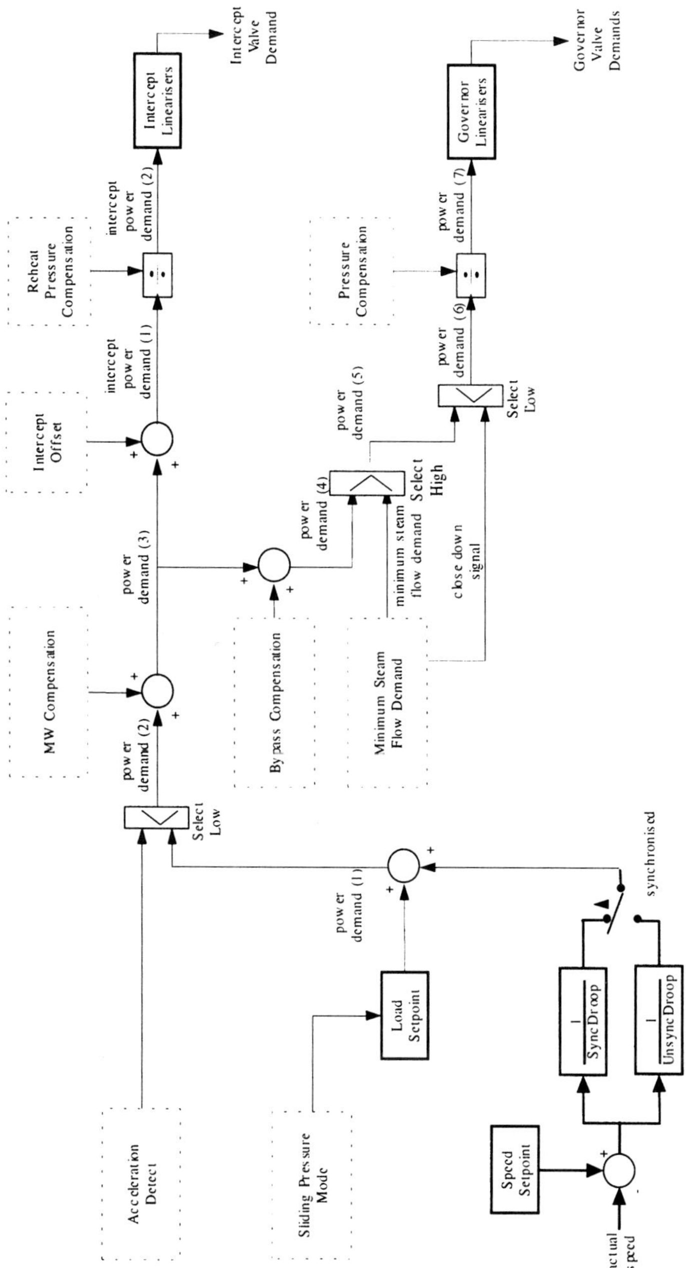

Figure 5 - EHG with Bypass Control Schematic

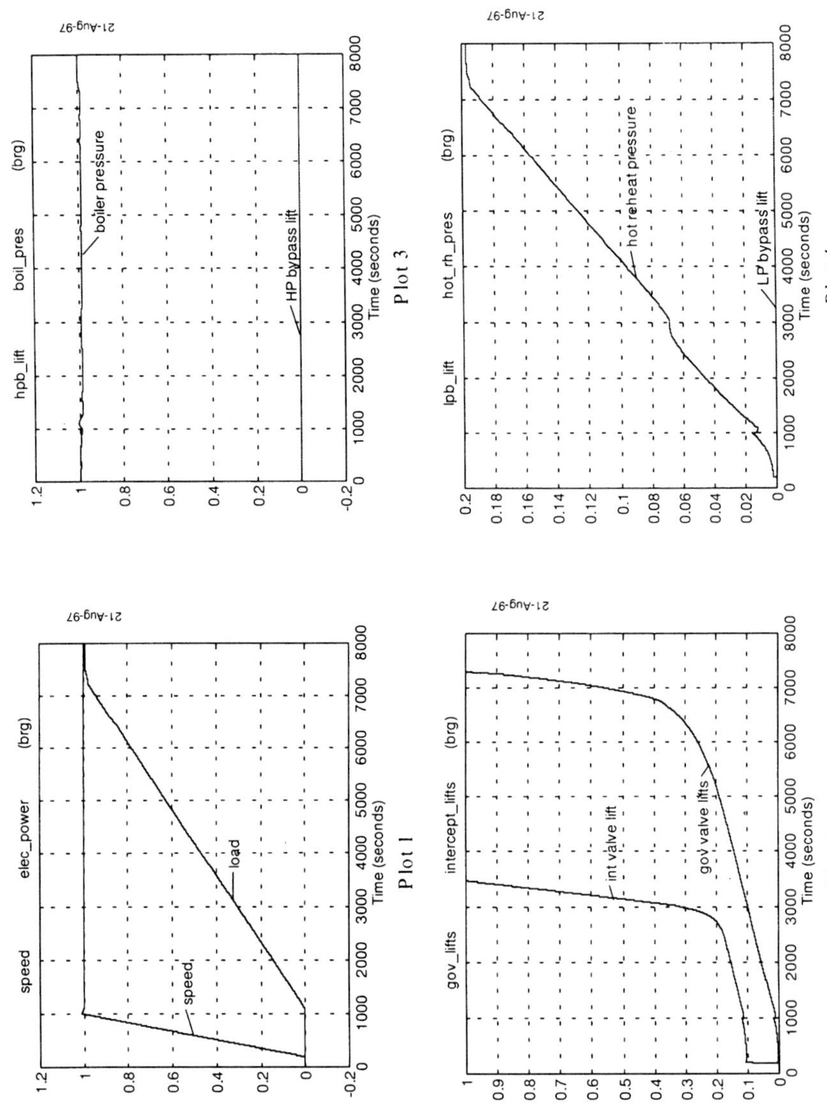

Figure 6 - Graph plots 1 to 4

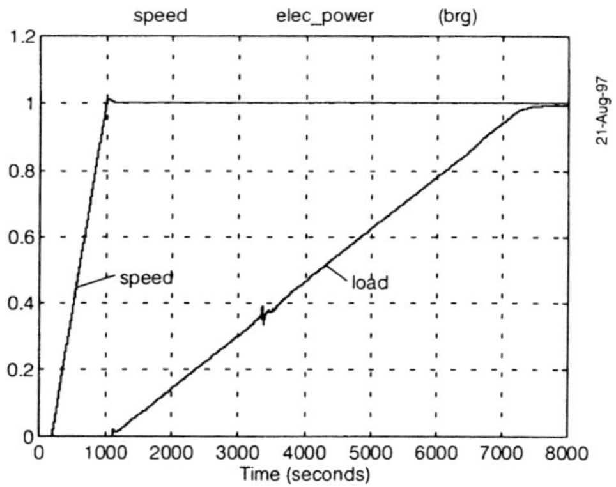

Plot 5

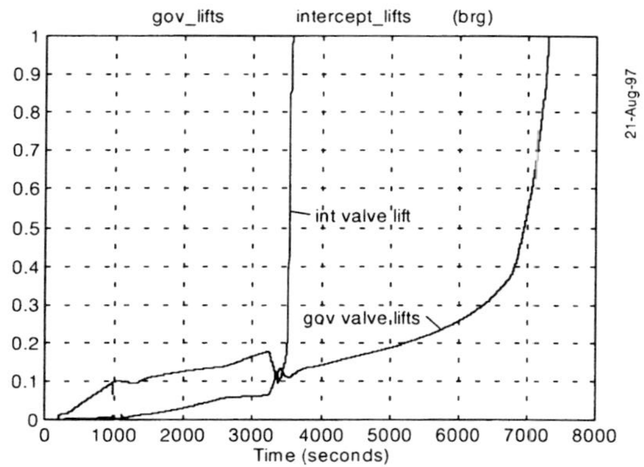

Plot 6

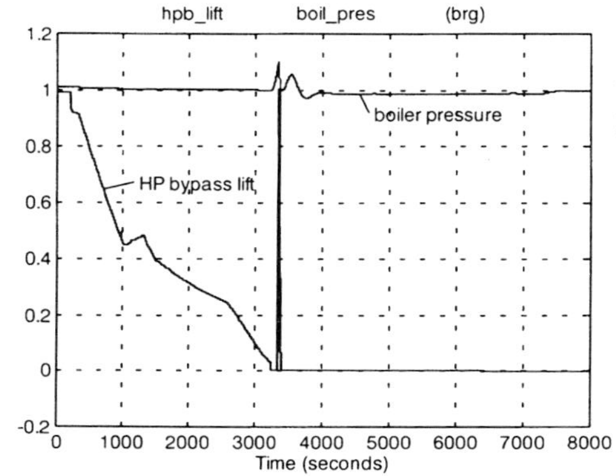

Plot 7

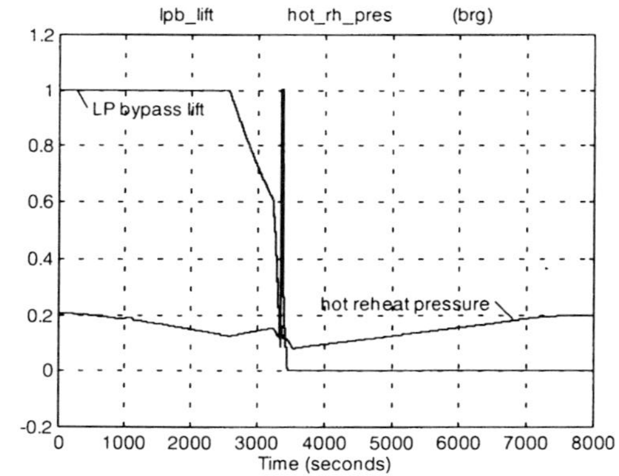

Plot 8

Figure 7 - Graph Plots 5 to 8

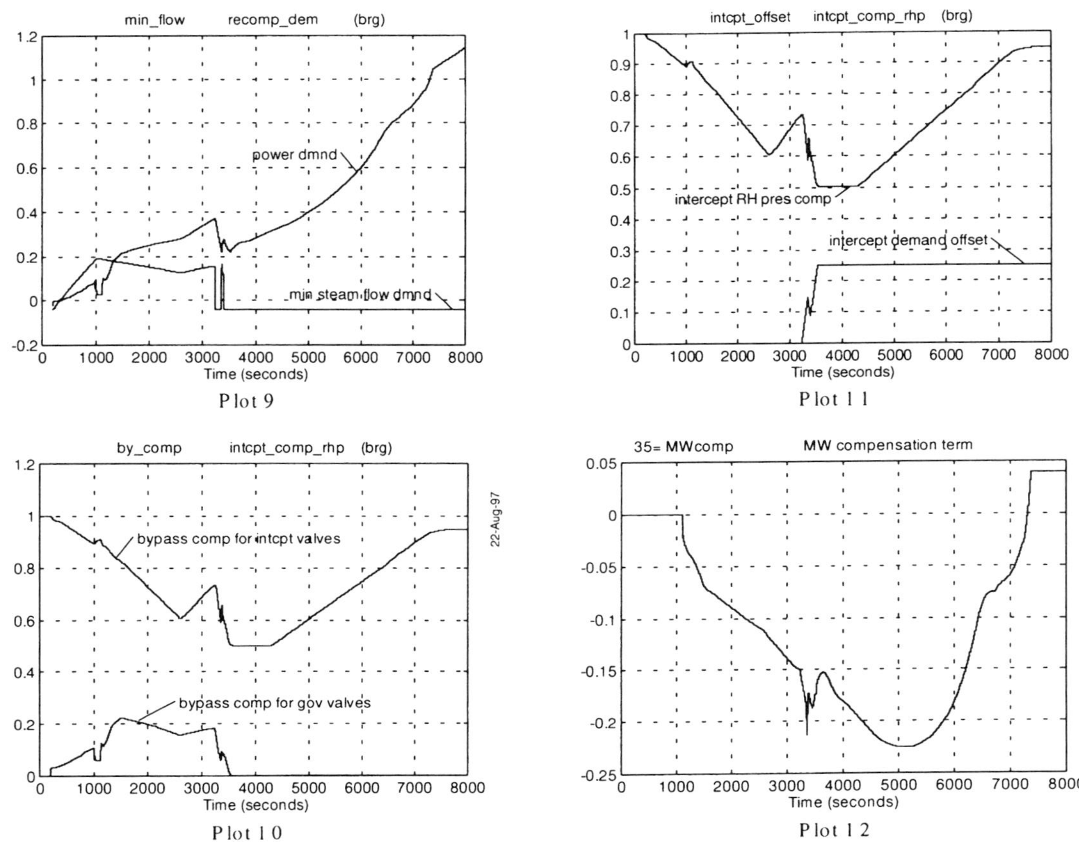

Plot 9

Plot 11

Plot 10

Plot 12

Figure 8 - Graph Plots 9 to 12

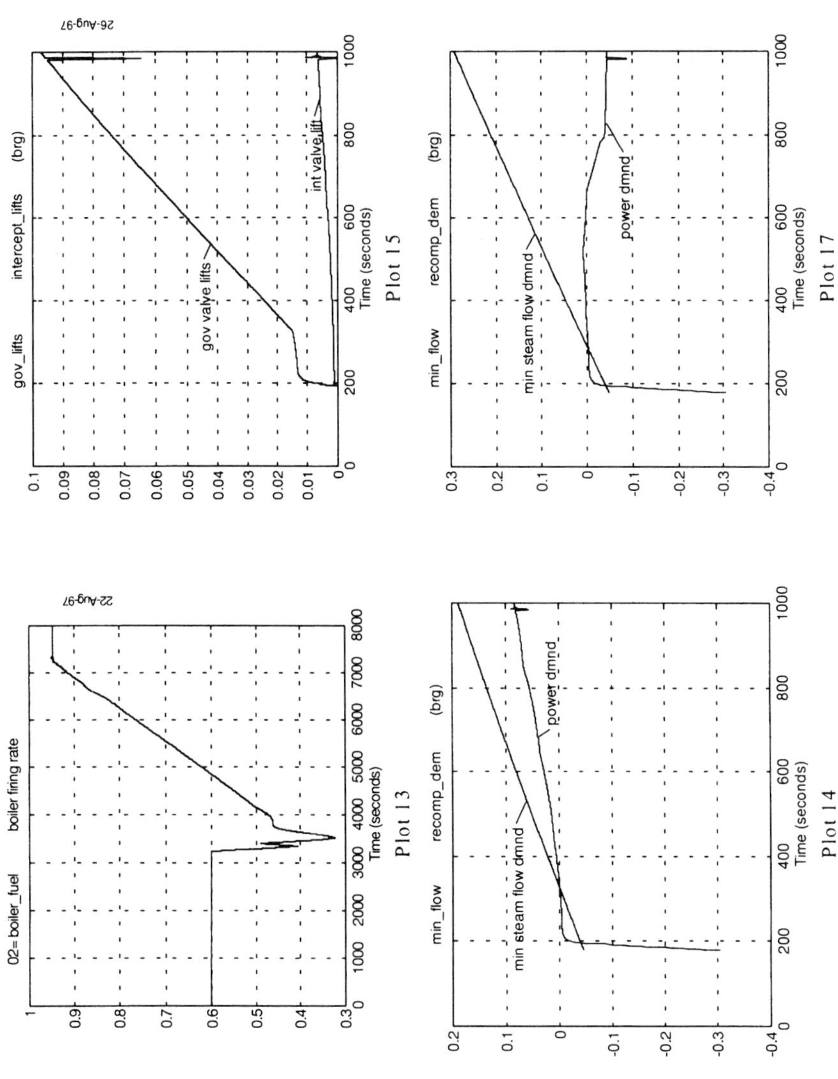

Figure 9 - Graph Plots 13, 14, 15 & 17

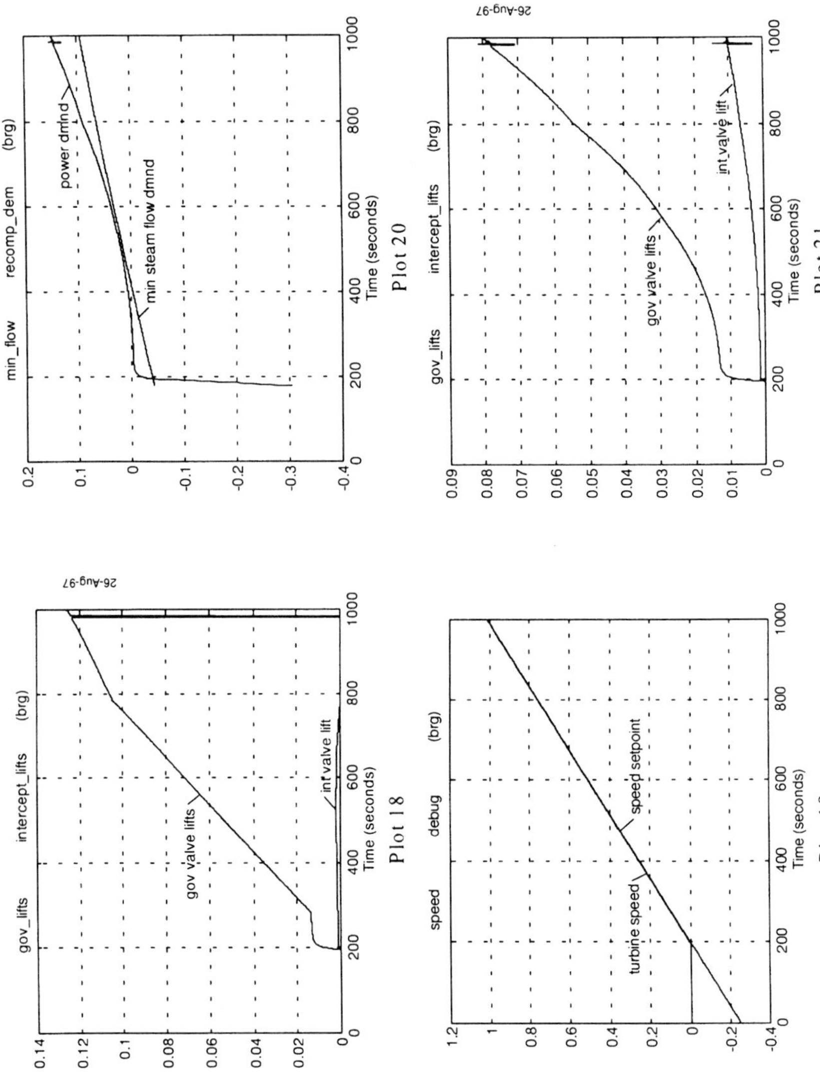

Figure 10 - Graph Plots 18 to 21

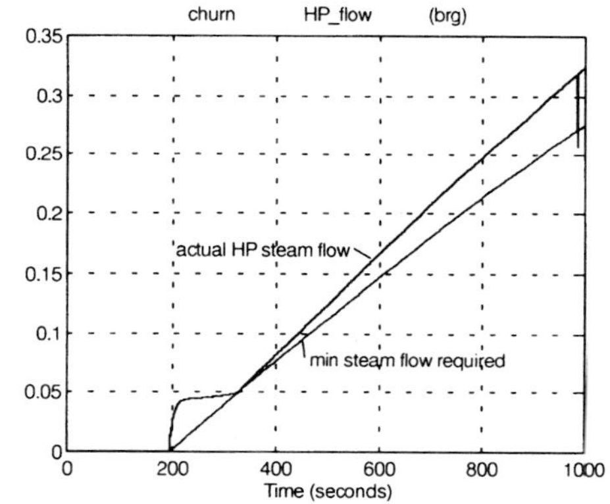

Plot 23

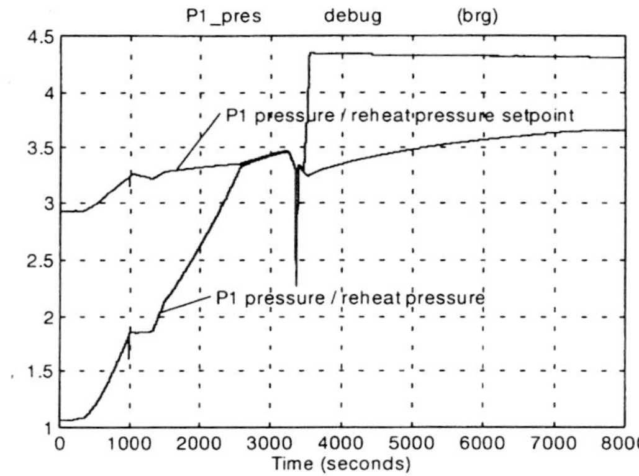

Plot 25

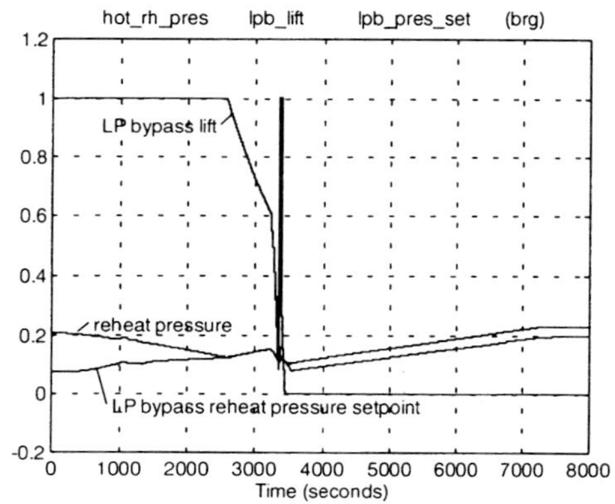

Plot 24

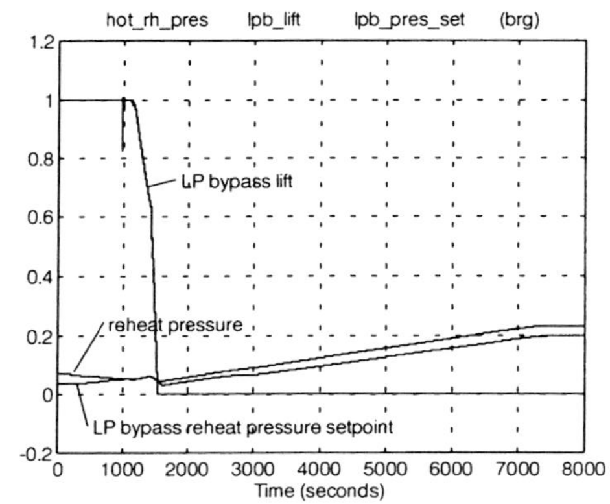

Plot 26

Figure 11 - Graph Plots 23 to 26

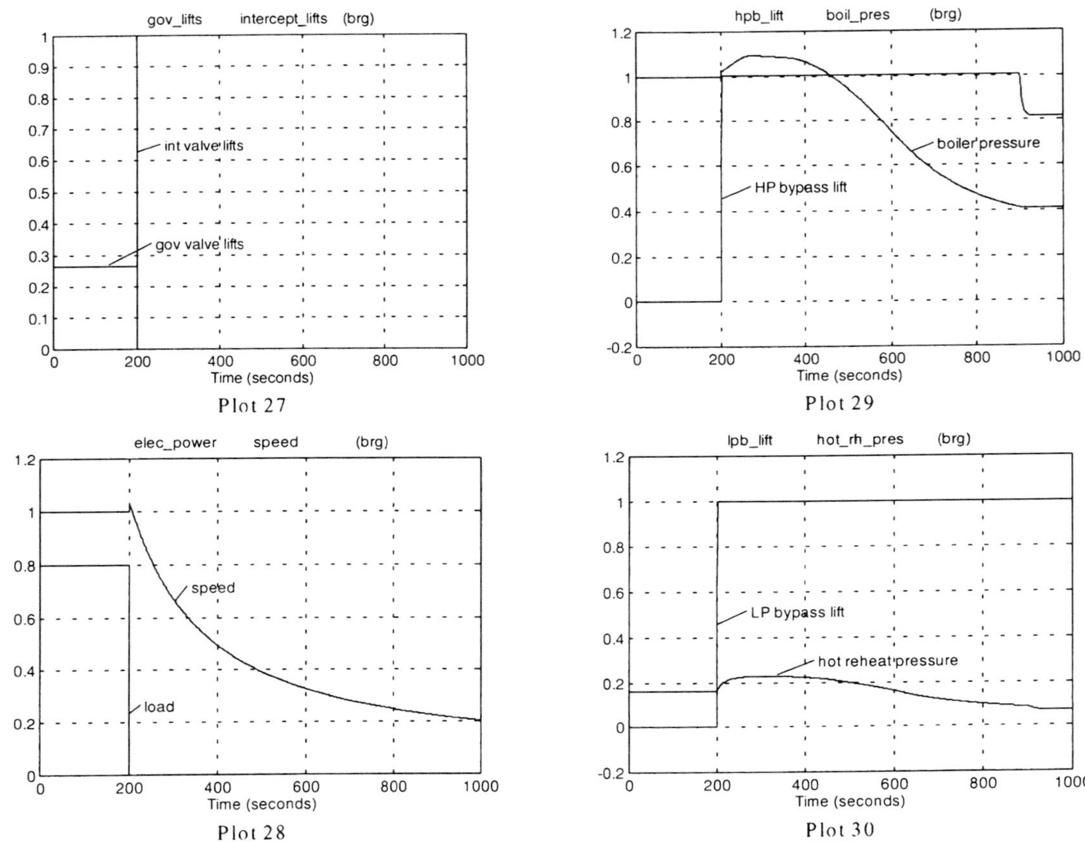

Figure 12 - Graph Plots 27 to 30

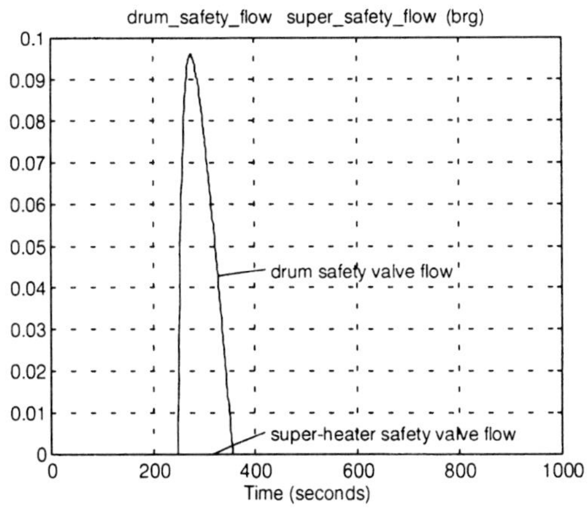

Plot 31

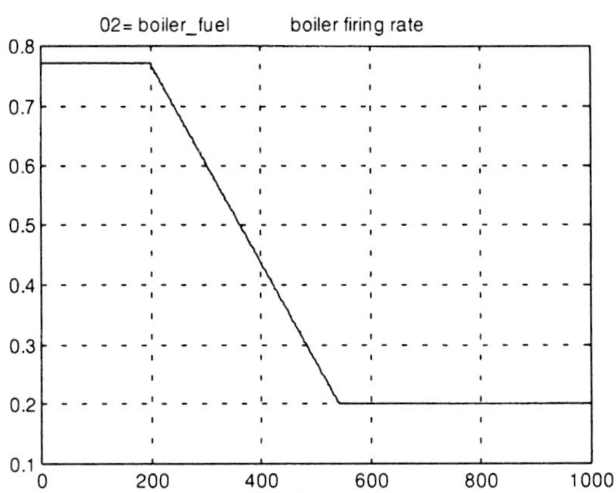

Plot 32

Figure 13 - Graph Plots 31 & 32

S593/006/98

CFD analysis of High Marnham Power Station condenser

R RHODES
Mott MacDonald, Croydon, UK
R PENNELLS and **R GHENT**
Eastern Generation Limited, UK

1 INTRODUCTION

Steam condensers form an integral part of a power station cycle. They provide a reduced back-pressure on the turbine by virtue of the condensation process and the provision of air pumps which maintain a pressure typically in the range 50 to 90 mbar. This low pressure increases the efficiency of the cycle, and it has been the challenge of condenser designers to optimise the placement of tube surface, drip trays and baffles in order to avoid high pressure drop through the condenser. With the advent of three-dimensional flow simulation techniques it has become possible to revisit past designs and consider modifications which improve the performance.

The overall performance of the condenser is affected by the arrangement of the tube nest and baffles within the shell. The low pressure in the condenser induces some air leakage through turbine glands and joints in the set. The successful removal of this air is a primary requirement in order not to impair the heat transfer performance by air blanketing effects which would increase the pressure.

The design process of most condensers has followed a fairly traditional engineering route. A design "code" gives a target heat transfer coefficient from which a total surface area is calculated, accounting for local cooling water temperature values peculiar to the particular site and the heat load and quantity of steam which the condenser has to handle. These parameters are easily established. The arrangement of the tube nest to promote good air removal and low pressure drop is not so straightforward. The difficulty in predicting how a particular tube nest design will perform in this respect has led to a wide variety of designs.

The use of Computational Fluid Dynamics (CFD) provides an alternative way of assessing the complicated physical processes that occur in a condenser. In these techniques a mathematical model is developed to represent the geometry of the condenser and the turbine exhaust region, dividing this volume into many smaller volumes or cells. In each cell calculations are performed to predict the flow velocities, pressure, concentration of the steam and air mixture and, in cells containing tubes, the cooling water temperature. Empirical correlations are used to calculate pressure drop, heat and mass transfer and condensate inundation, but they are applied independently in each grid cell, so accounting for local conditions.

The regions in the calculation domain which contain the tube nest are represented using a porous media approach. Pressure drop and condensation are calculated in these regions using empirical correlations. The cooling water temperature is also calculated. Baffles can be represented as blocked cell faces, and steam lanes and turbine exhaust regions are characterised by the absence of tubes, but may contain baffles and equipment items such as feed water heaters. This general approach has been developed over a number of years and has resulted in a fairly refined model with a high degree of validation. References 1 and 2 describe the details of the model application methodology and some early validation work.

The application of CFD methods to predict condenser performance have been carried out to for High Marnham Power Station's condensers with the purpose of investigating internal changes which might lead to improvements. The station has five identical turbine and condenser units with a total output of 1000MW. All of the units operate above their design target back pressure and, despite their nominally identical construction, one of the units operated noticeably further from its target. To investigate this situation inspections of the condenser were carried out and a CFD model of the present configuration was developed. Results from simulations of the "as-built" configuration indicated that steam bypassing directly to the air offtake was a possible cause, and this was later confirmed by additional measurements. One of the consequences of the high level of steam bypassing was a large loss of condensate through the air pumps and the need to run all of the available air pump capacity.

Following this verification, the model was used to explore various options for improving performance, and these have now been implemented in the plant. This paper describes the study and the results obtained.

2 PHYSICAL SITUATION

The condensers at High Marnham are of single vacuum, two-pass type. The 70-30 brass tubes are arranged in two separate bundles allowing water to enter in the lower sections and to be returned through the condenser in the upper sections. There are 17200 tubes in total, arranged in a triangular pitch. The pitch and diameter are 36.51 mm and 25.4 mm respectively. The tubes have 18 SWG wall thickness and are cleaned internally using a ball cleaning system. The cooling water enters the condenser via two inlet pipes connected to the lower water boxes. It returns through the upper nest section to two outlet pipes. Figure 1 shows a cross-section of one half of the condenser shell and tube nest. The condenser is symmetrical about its centre-line.

Internally, the condenser contains a number of baffles that direct the steam flow and a baffle on either side of the shell that forms the air cooling sections. In the early history of the sets holes were cut in the baffles marked 'D' and 'E' in Figure 1 to improve the flow through the condenser. The air-cooling section baffles cover the tubes in front of five air offtake pipes which run along each side of the condenser. These are connected through a manifold to two Nash pumps and three LeBlanc rotary air pumps.

The table below presents the performance of the five units for a cooling water temperature of 15°C and at full load conditions.

Unit number	Unit 1	Unit 2	Unit 3	Unit 4	Unit 5
Back Pressure (mbar)	44.3	48.8	61.4	48.3	43.3
Target Back Pressure (mbar)	42.3	42.3	42.3	42.3	42.3
Deviation from target (mbar)	2	6.5	19.1	6	1

Table 1: Summary of Performance Data

It can be seen that Units 1 and 5 are the best performers, 2 and 4 are similar and slightly poorer and Unit 3 is noticeably worse. The orientation of the units in terms of the cooling water system is such that 1 and 5 are on the outside ends of the circuits and Unit 3 in the middle. This might imply a maldistribution of water supply. However, calculations for the system indicated that variation in supply might only account for 1 mbar difference between Units 1 and 3, and that considerable differences in supply would be required to create the observed difference. This was therefore ruled out as a possible cause of the performance imbalance.

Two other items of data were of importance in understanding the system behaviour. Firstly, all five air pumps were in operation on each set and the effect on condenser back pressure of switching off the air pumps in different combinations was not the same. Figure 2 shows this effect by plotting the back pressure increase when different air off-take pumps are switched out. It can be seen that there is little effect initially, when only one Rotary Air Pump (RAP) is switched off, but the effect on back pressure is more marked as more air pumps are taken out of service. In addition, the increase in pressure is more marked for Units 1 and 5, the better performing sets, and less marked for Unit 3. Secondly, the variation in pressure with load was similar for all Units except Unit 3. This is shown in Figure 3 and it is clear that the pressure rises more steeply with load in Unit 3 than in the other Units.

3 MATHEMATICAL MODEL

A three-dimensional model has been used to predict the condenser performance on a grid which has been distorted to accommodate various features of the tube nest. A grid size of 61 x 40 cells has been employed in the cross-section of the condenser, and 20 cells in the direction of cooling water flow.

The equations solved in the model represent the conservation of momentum, steam concentration and mass continuity, and have the general form:

$$\Delta.(\rho U\phi - \Gamma_\phi \Delta\phi) = S_\phi$$

Where Φ represents the variable in question, Γ_Φ the diffusion coefficient and S_Φ the source or sink term.

Equations of the above form are solved for the velocities in the three co-ordinate directions, steam concentration and pressure. It is assumed that the steam is at saturation temperature corresponding to its local pressure. An enthalpy equation is therefore not required. The source terms in the equations include the effects of heat- and mass-transfer and frictional pressure drop. These are calculated within each grid cell containing tubes using empirical correlations which utilise local values of the variables.

Heat and mass transfer are calculated using a method described by Chisholm (3). It is assumed that the heat flux, q, and mass transfer m, are related through the latent heat by:

$$\dot{m} = \frac{q}{\lambda}$$

The heat flux is a function of the local temperatures of the vapour, condensate, outer tube wall and cooling water, T_v, T_c, T_w, and T_l respectively, and the corresponding heat transfer coefficients between cooling water and outer tube wall a_{wl}, across the condensate film, a_c, and the overall value between vapour and liquid, a_{ov}. The mass transfer can be deduced from the temperature gradient between vapour and condensate surface.

Tube friction is calculated assuming turbulent cross-flow conditions. The source terms have the form:

$$F = 2uf\rho N$$

where:

$$f = (0.23 + \frac{0.11}{(P/d_o - 1)^{1.08}})\,\mathrm{Re}^{-0.15}$$

and N is the number of tubes per cell.

4 PRESENTATION OF RESULTS

The results are presented in Figures 4 to 6 in the form of velocity vector plots and contour plots of steam concentration and pressure. Figure 4, the vector plot, shows the general flow into the condenser. Steam and air mixture enters at the top and flows into the upper tube nest, accelerating into the steam lanes between the tube nest sections. A proportion of the steam passes down the central lane, ie at the symmetry plane on the right of the figure, to the floor of the condenser. It then flows upwards into the mid-lane of the lower tube nest. The prediction indicates that this flow path supplies most of the bottom section with steam, and relatively little flows from the top section.

The steam-air concentration is shown in Figure 5, where it can be seen that the upper tube nest has a high concentration of steam, whilst the lower section has extensive air blanketing on both sides. This indicates that most of the condensation for these flow conditions occurs in the upper part of the nest and the lower part is under utilised. This figure also shows a high air concentration at the air off-take point. The prediction indicates that this is caused by the central flow which passes along the floor of the condenser passes baffle B. Although the steam has to pass through the tube nest, there would appear to be insufficient surface exposed to steam to ensure full condensation. This will have the effect of swamping the air offtake pumps and causing the pressure in the condenser to rise.

The pressure field in the condenser is shown in Figure 6. The main feature that this shows is the high pressure-gradient in the upper tube nest. In the lower part the pressure is relatively uniform. Some pressure recovery can be seen when the central flow reaches the floor.

Comparing measured pressure data with prediction for full load conditions the predicted pressure was 52.2 mbar. Comparing with the measured data in Table 1, it can be seen that this is between the best and worst performing condensers, suggesting that the model is a reasonable average representation of the sets. The prediction of steam bypassing to the air off-take led to a recommendation to secure measurements of the steam temperature and pressure at the offtake. Table 2 shows the results of these measurements, which confirmed that steam offtake temperatures were well above the cooling water temperature.

Unit	CW Inlet Temperature (°C)	Air Off-Take Temperature (°C)	Air Off-take Pressure (mbar)	Condenser Pressure (mbar)
1	18.8	31.5	55.0	57.2
3	17.3	33.4	65.0	65.0

Table 2: Steam Off-Take Conditions

Further studies with the model were therefore undertaken to evaluate the effect of suppressing steam bypassing. The modelling demonstrated that it should be possible to reduce the condenser pressure by about 5 mbar. The station then tested a modification which involved extending baffle 'B' into the nest by using neoprene sheet, bolted to the existing baffle and pushed between the tubes. This reduced the extreme off-take conditions and brought about an improvement in some of the units. Table 3 indicates the effect.

Unit	CW Inlet Temperature (°C)	Air Off-Take Temperature (°C)	Air Off-Take Pressure (mbar)	Condenser Pressure (mbar)
1	19.2	23.4	39.1	45.6
2	24.1	26.8	50.0	54.8
3	13.9	21.5	32.5	36.7
4	19.5	26.8	44.7	48.2
5	20.3	25.9	44.1	47.8

Table 3: Typical Steam Off-Take Conditions After Modification

In addition to the extension of baffle 'B' on all units, baffles 'D' and 'E' were removed from Unit 3 and the 700 plugged tubes were replaced. The overall improvements in back-pressure were as follows:

Unit	Back Pressure Improvement (mbar)	Water Loss Improvement (t/h)
1	0	3.2
2	5	2.5
3	10	4.0
4	5	3.1
5	5	2.9

Table 4: Overall Improvements in Back Pressure and Water Loss

The improvement in back-pressure has resulted in an increase in cycle thermal efficiency and a consequent reduction in fuel costs worth £157,000 per annum at a station load factor of 30%. There is an additional saving of 53,000 tonnes of condenser make-up water per year at the same station load factor, giving a further saving of £4,400. The modification work had a very short pay-back period and as soon as the success of the first modified unit had been realised, High Marnham implemented the modification on the other units at the first opportunity.

The effect of removing some of the air extraction pumps from operation has been examined on Units 2 and 3, and it has been found that the effect on condenser back pressure is now similar for both condensers. When one Nash pump was switched off the back pressure rose by 0.6 and 0.7 mbar respectively in Units 2 and 3. Switching out the second Nash pump increased these to 2.8 and 4.6 mbar. Thus the modification work now allows the air pumps to work effectively as steam is no longer by-passing to the air off-take.

5 CONCLUDING REMARKS

A study has been carried out to understand the detailed flow and performance of the condensers at High Marnham Power Station. The development and application of a three-dimensional CFD model has provided a means of investigating some aspects of the performance of the sets and bringing about valuable improvements in performance. The improvements in back pressure on four of the five units have led to reduced operating costs and less condensate is lost through the air pumps.

6 REFERENCES

1. Else, K and Jennings J J "Mott MacDonald Steam Condenser Model User Manual" (1990)

2. Al-Sanea S, Rhodes N, Tatchell T G and Wilkinson T S " A Computer Model for Detailed Calculation of the Flow in Power Station Condensers". IMechE Symposium Series no. 75, pp 70-88. (1983)

3. Chisholm, D "Modern Developments in Marine Condensers: Non-condensable Gases: An Overview" **Power Condenser Heat Transfer Technology**, edited by Marto and Nunn, pub. by McGraw-Hill (1980).

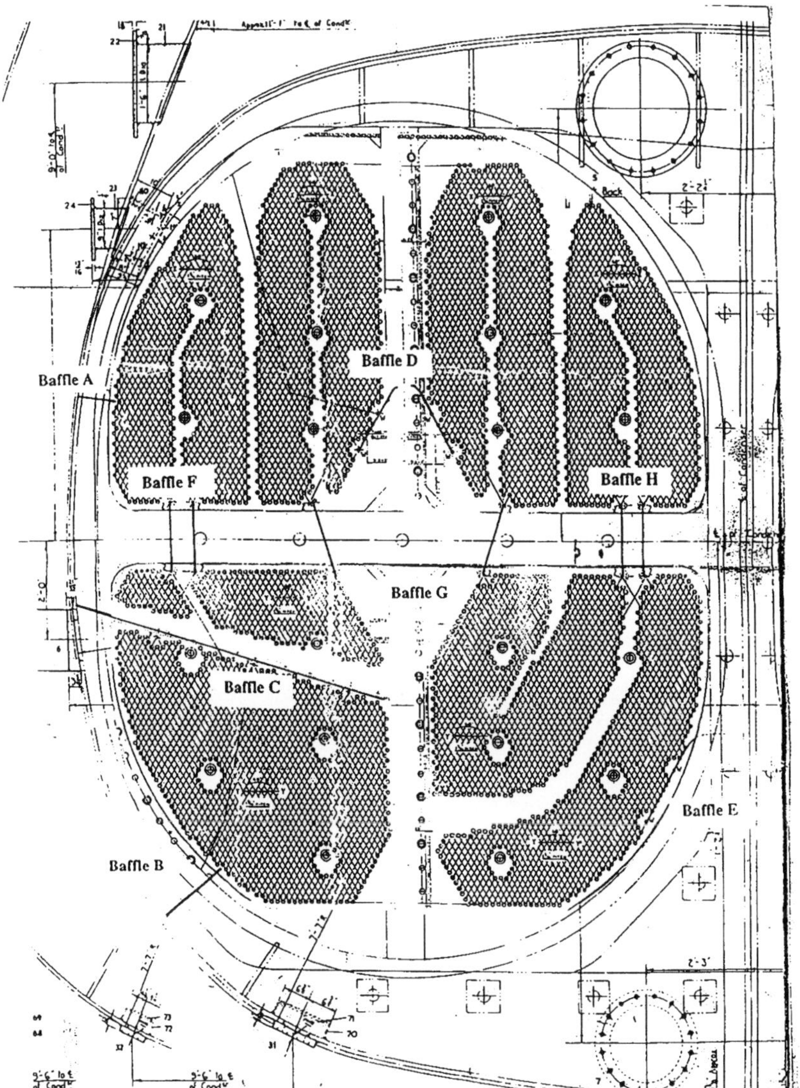

Figure 1: Tube Nest Geometry

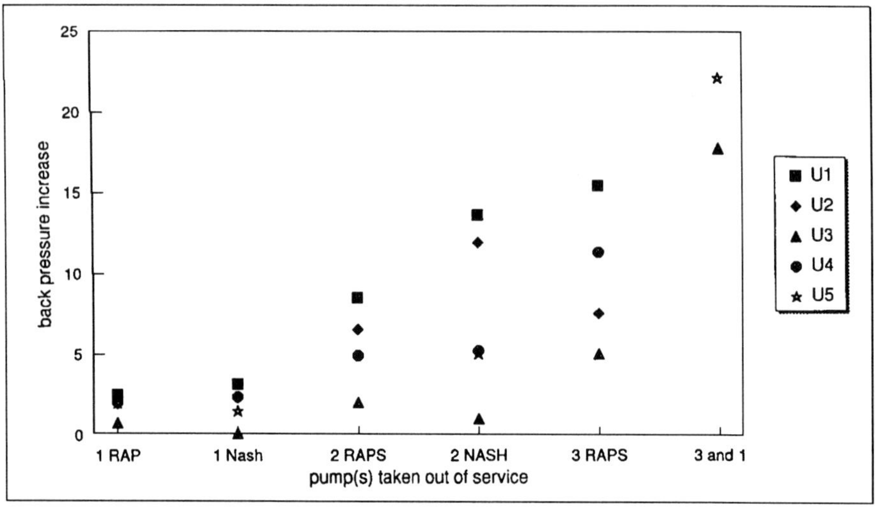

Figure 2: Graph of Condenser Back Pressure Increase versus Pump or Pumps Removed from Service
(Two Nash pumps and three Le Blanc Rotary Air Pumps normally running)

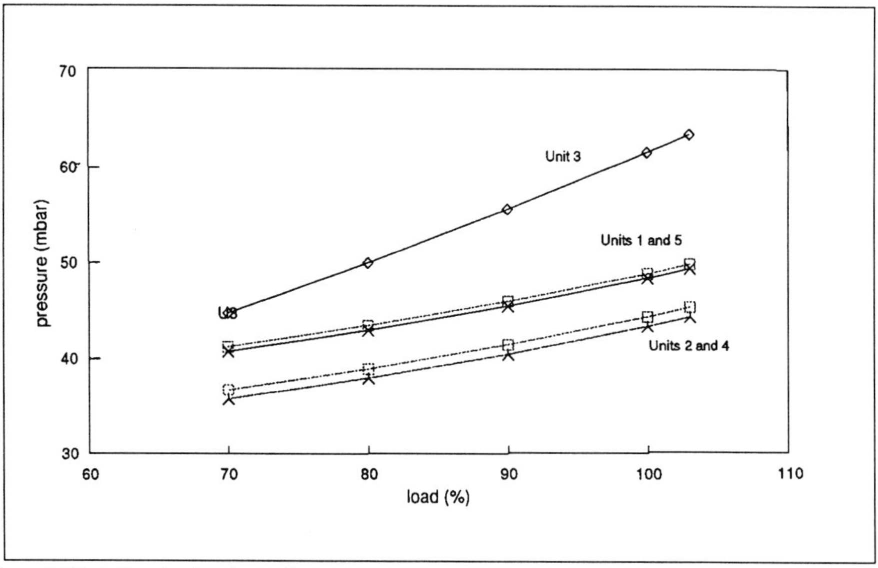

Figure 3: Effect of Load on Condenser Pressure

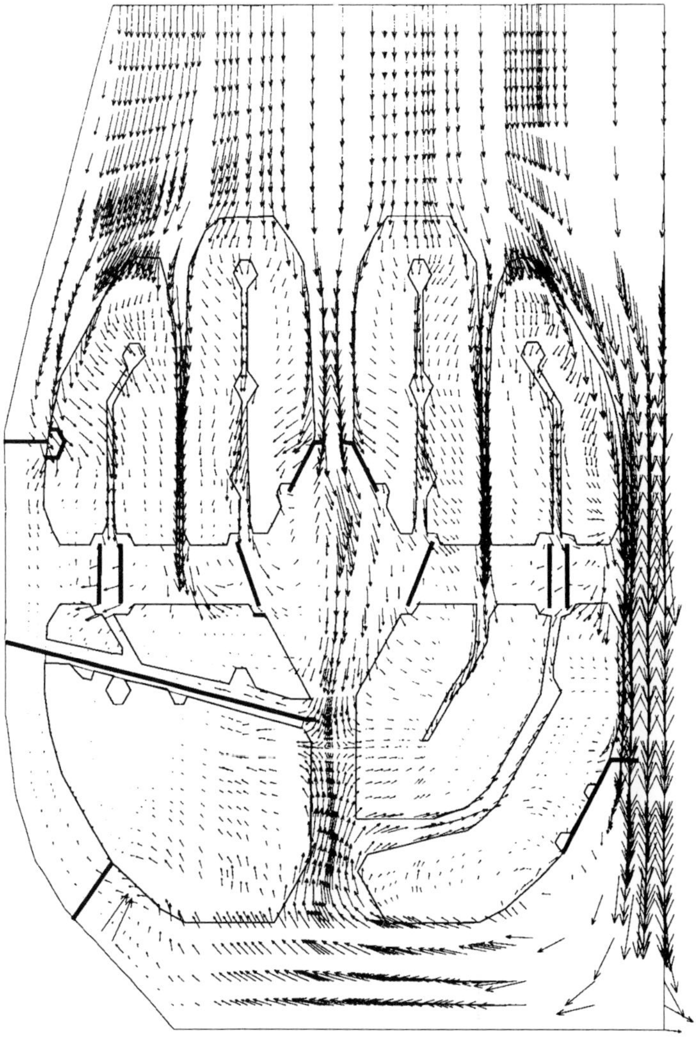

Figure 4: Vector Plot

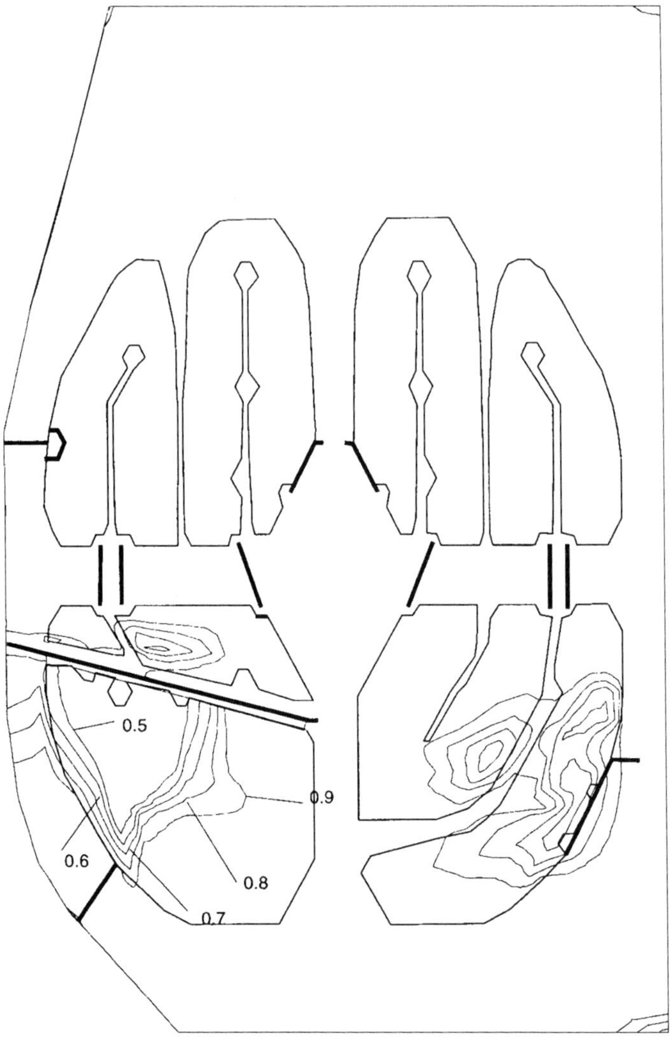

Figure 5: Steam Concentration Contours

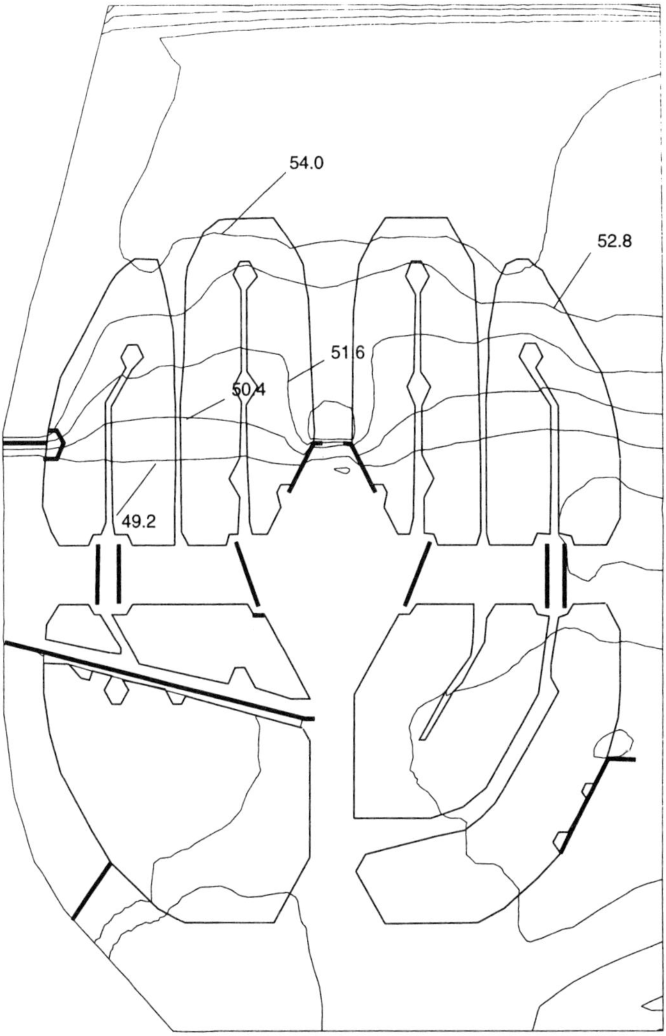

Figure 6: Pressure Contours (mbars)

S593/007/98

Computational fluid dynamics for flow visualization in steam turbine inlets and exhausts

W G HORSEMAN
Siemens Power Generation Limited, Newcastle-upon-Tyne, UK

SYNOPSIS

Application of Computational Fluid Dynamics methods in steam turbine inlet and exhaust design can give an early appreciation of the flow behaviour. The solution of complex three-dimensional flow fields is always challenging, and it is important that attention should be given to the resolution of influential flow features through the use of properly formulated models. The knowledge gained through an understanding of the flow can be used to reduce the need for experimental investigation, which helps to shorten design lead-times and to improve thc use of development resources.

1. INTRODUCTION

Computational Fluid Dynamics (CFD) methods are being used increasingly as a principal analytical tool in the design of large-scale steam turbines. Historically, the difficulty of access and of direct steam flow measurement resulted in design principles being derived from practical experience and the use of scale models. Measurement technology is mature and produces reliable information, but direct observation of flow behaviour is difficult. Design improvement by this means may involve an expensive iterative exercise. CFD methods are being used in practical turbine design, and experimentation is being used mainly for verification. One benefit is that CFD methods give an improved appreciation of the flow, which leads to quicker identification of the controlling factors. Early recognition of flow behaviour improves the cost effectiveness of the development process. For turbine inlets and exhausts where the flow is substantially three-dimensional, CFD techniques are less well established and experimental support has been needed.

The complexities of three-dimensional flows have given rise to intense academic interest and development. Generalised CFD codes are available commercially and with the improving robustness of 3-D Navier-Stokes solvers, there is now widespread use of viscous methods in the development of turbine technology. Although these flow regimes are complex, the constraint of design lead-time often precludes an academic approach to the modelling. Available techniques may be only partially suitable for the analysis of the flow regime and simplifications are required to bound calculation times. This paper is concerned with the methodology of applying available CFD techniques in turbine design in circumstances that preclude a long-term approach.

The subject of the paper is illustrated using two examples of three-dimensional turbulent flow. Both cases are related to existing designs with service experience and model test results to support the work. The designs operate with good efficiencies, but there may be scope for further improvement by the reduction of recirculating flow. One case represents a low-pressure turbine exhaust hood arrangement, and the other an IP inlet belt. Two different solution methods have been used, both employing commercial solvers. These examples have been chosen to validate the analytical techniques and so enable further design refinement.

2. MODELLING

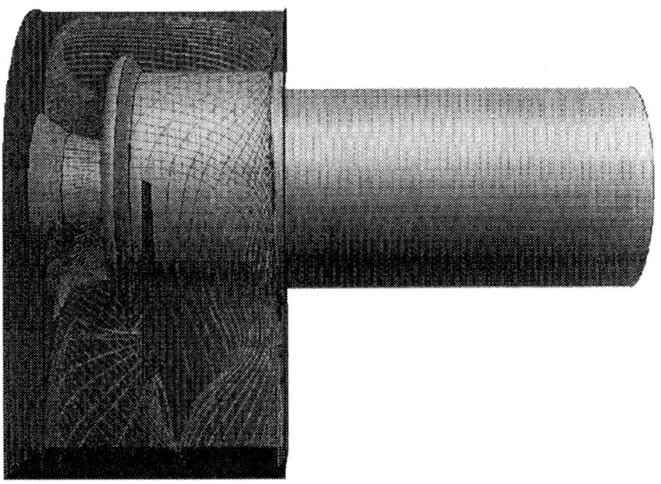

Fig. 1 Exhaust hood and diffuser

The computational models are derived from air driven test rigs which are scaled versions of the turbine components so that experimental results can be used to validate the analyses. For the purposes of the study a longitudinal plane of symmetry on the vertical axis of the turbine is assumed. This is strictly applicable only to the experimental models. Although the effect of swirl is relevant to the operational case, entry flows are normal to the inlet plane with no significant swirl component. In both cases the computational grids use a multiblock structured mesh, the physical boundaries of which are closely mapped using b-spline surfaces. Comparison of solutions showed that the residual mesh dependency is unlikely to obscure the engineering interpretation. Some refinement of the grid construction was carried out in order to minimise the occurrence of severely skewed elements. The intention was to establish computational models, which could be evaluated with reasonable turnround time. For this reason a conventional two-equation eddy viscosity approach is taken for the turbulence modelling. However, it is noted that the effect of turbulence modelling is important and some deficiencies of the model attributable to the turbulence are apparent.

2.1 Case 1 LP Exhaust Hood

The computational model represents a 1:30 scale experimental rig operating at realistic mach numbers in the transonic region. It has a single vertical symmetry plane and so includes 180^0 of the diffuser circumference. The inlet plane is at one end of a cylindrical inlet duct upstream of the diffuser. A bullet nose is used to separate the axial inflow into an annulus and the exit plane of the model is at the exhaust flange plane. The steam inlet pipe is omitted together with various minor details of the hood and turbine casing and all internal struts and ribs. The multiblock structured mesh consists of approximately 100 000 fluid cells.

The solution of the Reynolds averaged Navier-Stokes (RANS) governing integral equations was obtained using a time marching explicit numerical scheme of nominal second order accuracy with multigrid convergence. Spatial discretisation is based on a finite volume representation. Turbulence modelling in this example uses a standard k-epsilon two-equation approach with equilibrium wall functions. The walls are adiabatic and non-slip. In order to improve solution times implicit residual smoothing is used. Smoothing is applied using a Laplacian operator over two iterations. The model is closely matched to the diffuser profiles, although due to the use of a fully 3-dimensional model, there is some sparseness in the grid near the outer diffuser boundary and hood. This illustration is chiefly concerned with flow near the inner hub boundary, for which the mesh density appeared reasonable in accordance with the principles of the wall function approach.

2.2 Case 2 IP Inlet Belt

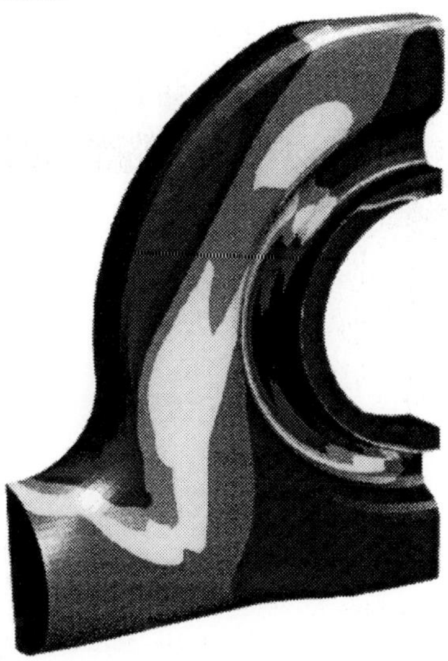

Fig. 2 IP Inlet belt showing total pressure surface contours

The inlet belt is modelled from an experimental arrangement at quarter scale. Incompressible, isothermal flow and adiabatic walls are assumed. The model has two planes of symmetry in order to represent a double flow IP turbine arrangement with dual inlets. The physical model is also single ended and uses solid plane surfaces for the same purpose. The inlet pipes are cylindrical and of a nominal length and are not intended to represent flow through the corresponding steam chest.

The solution method is described as a finite element scheme using a conservative finite volume mesh. The governing equations use eddy viscosity (two-equation) turbulence closures, as above, but in this case the dependence on turbulence modelling is illustrated for two cases, a standard k-epsilon model and a Renormalisation Group derived k-epsilon (RNG) model. The mesh was constructed with approximately 30 domains and contains over 200 000 nodes. Discretisation is by linear profile skew upwinded differencing with physical advection correction. This is a second order method, but contains diffusion terms which can be destabilising. Consistency checks were carried out using blended schemes of lower order to demonstrate reasonable solution independence.

Downstream of the duct, the bladepath resistance was represented by a simulated porous screen as in the experimental model. In order to avoid severe gradients in the flow the porosity was introduced with substantial length and the resistance varied to match empirical data.

3 CFD EVALUATIONS

3.1 Case 1 LP Exhaust Duct

Flow behaviour near the walls of the diffuser is of primary importance to the design. On the hub, separation may occur at the shallower section of the bearing cone, but the recirculation region is confined close to the hub and has been shown in practice not to cause any major obstruction to the flow.

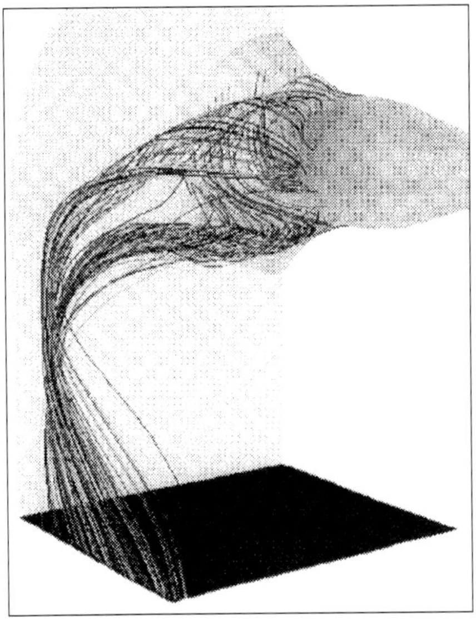

Fig. 3 Flow dispersal from hub

Fig. 4 Flow features on underside of hub

As a visualisation tool, the CFD method can give immediate access to the flow in order to determine the main features. The complexity of the flow at the hub suggested the presence of a number of separation and recirculation features. Within the primary recirculation, there is circumferential migration from the uppermost and lowermost surfaces of the hub to a position near the horizontal axis. The principal flow is swept up into vortices, which detach from the hub surface and impinge the wall of the outer cylinder. This can be tracked as seen in figure 3.

Comparison with experiment is made by inspection of flow sensitive paint. Undisturbed paint indicates low velocity and the likelihood of flow dispersal. The regions of accumulated paint are attributed to impinging flows and low velocity flow adjacent to a higher flow regime. Migration of paint from upstream of the bearing cone to the shallower conical section demonstrates that the primary separation is in general accord with the CFD prediction. Large patches of accumulated paint near the horizontal axis and streak lines on the hub indicate the circumferential flow. Close inspection of the surface reveals four circular patches of undisturbed paint. Two of these below the hub are shown in figure 4. These represent the centres of vortices where the flow is detaching from the hub surface.

3.2 Case 2 IP Inlet Belt

The shape of the elbow at entry to the IP inlet belt influences the flow distribution. In this case the flow is known to separate from the elbow and recirculate in the upper half of the belt. This affects flow incidence at the blade leading edge. In the experimental model, traverses are made across the blade path to determine the velocity and pressure distribution upstream of the first stage blade and the variation in magnitude and direction with blade height.

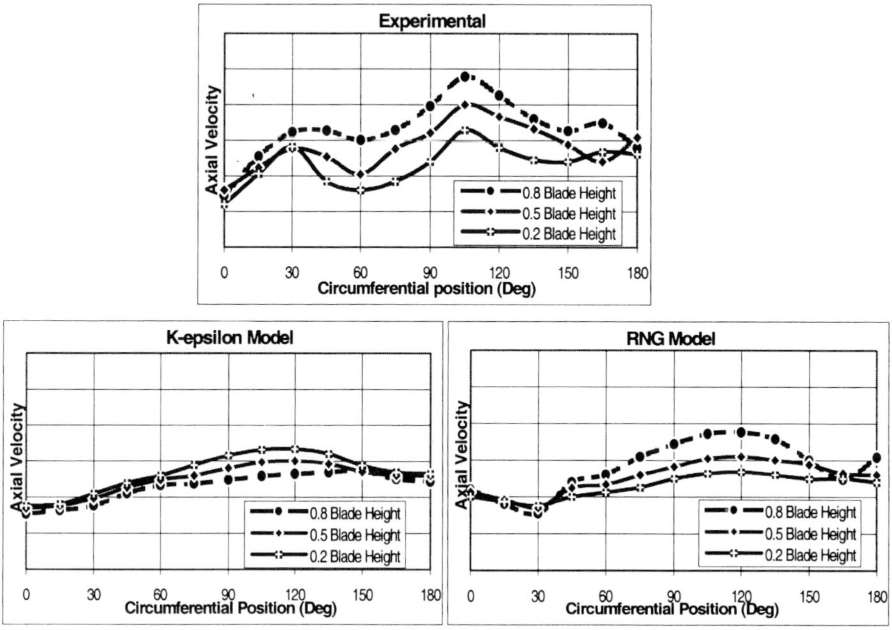

Fig. 5 Axial profile with blade height

Comparison of predicted velocity and incidence with measurement shows good overall agreement. Where numerical discrepancies do occur, a number of factors were seen to be influential. Of these, the most noteworthy is the distribution of axial flow across the height of the blade, which for the standard k-epsilon turbulence model is shown as having higher velocities near the blade root (fig. 5). A solution using an RNG model is in closer agreement with the observations with the higher velocities at the blade tip.

4. DISCUSSION

For the IP inlet model, the severe streamline curvature and local impingement within the blade passage are not resolved by terms in the standard k-epsilon equations. This limitation is known, for example Launder (1), and demonstrated for a two-dimensional U-duct by Monson (7). In this case a defect in the total pressure is also observed, as noted by Hutchinson (8) for a constant density calculation of a Stanitz elbow. Although Hutchinson shows an improvement using finer grids, strong grid dependency was not reproduced in this calculation and the defect was corrected using the RNG turbulence description (fig. 6).

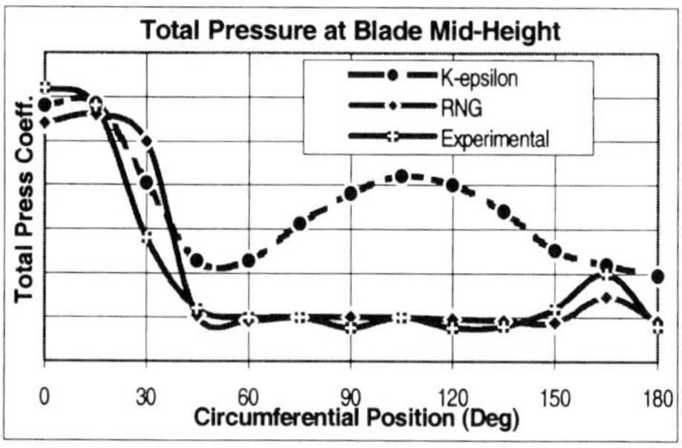

Fig. 6 Comparison of total pressure coefficient

In order to determine the validity of the model for design purposes, the calculation is shown to be capable of reproducing the major trends which influence the design. The principal recirculation features, flow magnitudes and blade incidence angles were all predicted sufficiently well. The flow distribution across the height of the blade was improved by use of a turbulence description with additional terms.

It is important to bear in mind that the reference design operates with good efficiency, in which these flow effects are not pronounced, so the ability to reproduce these effects in the analytical model shows that it can be used for design evaluation and refinement.

5. CONCLUSIONS

The given examples demonstrate the successful application of CFD methods to turbine inlet and exhaust development under real world constraints where lead-time and resources are important. The analytical results were able to reproduce the flow behaviour known to occur in operational machines with sufficient resolution to allow design optimisation. Care is needed to set up the analytical model, and some of the key factors such as discretisation of the flow field and the turbulence modelling have been discussed. The range of application of the CFD model is related to the available data, and there is a need for further experimental work if the analytical techniques are to be extended to more diverse applications.

6. ACKNOWLEDGEMENTS

The author gratefully acknowledges the support of his colleagues at Siemens, Newcastle and Muelheim for experimental data and technical information, in particular Mr M Carpenter, Mr M Wechsung and Dr M Deckers, also Mr P Hackwell at Rolls Royce, Ansty.

7. REFERENCES

1. Launder, B.E. Current Capabilities for Modelling Turbulence in Industrial Flows, J. of Appl. Scientific Research 48:247-269 (1991)

2. Yakhot, V. and Smith, L.M. The Renormalisation Group, The ε-Expansion and Derivation of Turbulence Models, J. of Sci. Comp, 3:35 (1992).

3. Boysan, H.F. Introduction to the Renormalisation Group Method and Turbulence Modelling. TM-107 Fluent Inc., Lebanon, NH. (1993).

4. Fluent-UNS and RAMPANT Users Guide 4.0. Fluent Inc., Lebanon, NH. (1996).

5. Raw M.J., Galpin, P.F. and Hutchinson, B.R. A Collocated Finite Volume Method for Solving the Navier-Stokes Equations for Incompressible and Compressible Flows in Turbo-machinery: Results and Applications, Canadian Aeronautics and Space J., 35:189-196 (1989).

6. TASCflow Theory Documentation, Advanced Scientific Computing Ltd., Waterloo, Ontario (1995).

7. Monson, H.L., Seegmiller, P.K., McConnaughey, Chen, Y.S. Comparison of Experiment with Calculations Using Curvature Corrected Zero and Two Equation Turbulence Models for a Two-Dimensional U-Duct. AIAA 90-1484 (1990).

8. Hutchinson, B.R., Elias S.R. and McMillan S.D. The Evaluation of a Numerical Error Estimation Technique for Engineering Flows. 4th Annual Conference CFD Soc. of Canada, Ottawa, Ontario (1996).

9. Hosenfeld, H. Stroehmungstechnische Modellversuche an einem MT-Einstroemgehaeuse mit Zweiseitiger Symmetrischer Zustroemung Unterhalb der Teilfuge. Siemens Internal Report (1993).

10. Raithby, G.D. Skew Upstream Differencing Schemes for Problems Involving Fluid Flow. Comp. Meth. Appl. Mech. Eng., 9:153-164 (1976).